Lecture Notes in Control and Information Sciences

Edited by M. Thoma

Vol. 22: Optimization Techniques
Proceedings of the 9th IFIP Conference on Optimization Techniques,
Warsaw, September 4–8, 1979
Part 1
Edited by K. Iracki, K. Malanowski, S. Walukiewicz
XVI, 569 pages. 1980

Vol. 23: Optimization Techniques
Proceedings of the 9th IFIP Conference on Optimization Techniques,
Warsaw, September 4-8, 1979
Part 2
Edited by K. Iracki, K. Malanowski, S. Walukiewicz
XV, 621 pages. 1980

Vol. 24: Methods and Applications in Adaptive Control
Proceedings of an International Symposium
Bochum, 1980
Edited by H. Unbehauen
VI, 309 pages. 1980

Vol. 25: Stochastic Differential Systems – Filtering and Control
Proceedings of the IFIP-WG7/1 Working Conference
Vilnius, Lithuania, USSR, Aug. 28 – Sept. 2, 1978
Edited by B. Grigelionis
X, 362 pages. 1980

Vol. 26: D. L. Iglehart, G. S. Shedler
Regenerative Simulation of Response Times in Networks of Queues
XII, 204 pages. 1980

Vol. 27: D. H. Jacobson, D. H. Martin, M. Pachter, T. Geveci
Extensions of Linear-Quadratic Control Theory
XI, 288 pages. 1980

Vol. 28: Analysis and Optimization of Systems
Proceedings of the Fourth International Conference on Analysis and Optimization of Systems
Versailles, December 16–19, 1980
Edited by A. Bensoussan and J. L. Lions
XIV, 999 pages. 1980

Vol. 29: M. Vidyasagar,
Input-Output Analysis of Large-Scale Interconnected Systems –
Decomposition, Well-Posedness and Stability
VI, 221 pages. 1981

Vol. 30: Optimization and Optimal Control
Proceedings of a Conference Held at Oberwolfach, March 16–22, 1980
Edited by A. Auslender, W. Oettli, and J. Stoer
VIII, 254 pages. 1981

Vol. 31: Berc Rustem
Projection Methods in Constrained Optimisation and Applications to Optimal Policy Decisions
XV, 315 pages. 1981

Vol. 32: Tsuyoshi Matsuo,
Realization Theory of Continuous-Time Dynamical Systems
VI, 329 pages, 1981

Vol. 33: Peter Dransfield
Hydraulic Control Systems –
Design and Analysis of Their Dynamics
VII, 227 pages, 1981

Vol. 34: H.W. Knobloch
Higher Order Necessary Conditions in Optimal Control Theory
V, 173 pages, 1981

Vol. 35: Global Modelling
Proceedings of the IFIP-WG 7/1 Working Conference Dubrovnik, Yugoslavia,
Sept. 1–5, 1980
Edited by S. Krčevinac
VIII, 232 pages, 1981

Vol. 36: Stochastic Differential Systems
Proceedings of the 3rd IFIP-WG 7/1 Working Conference
Visegrád, Hungary, Sept. 15–20, 1980
Edited by M. Arató, D. Vermes, A.V. Balakrishnan
VI, 238 pages, 1981

Vol. 37: Rüdiger Schmidt
Advances in Nonlinear Parameter Optimization
VI, 159 pages, 1982

Vol. 38: System Modeling and Optimization
Proceedings of the 10th IFIP Conference
New York City, USA, Aug. 31 – Sept. 4, 1981
Edited by R.F. Drenick and F. Kozin
XI, 894 pages. 1982

Vol. 39: Feedback Control of Linear and Nonlinear Systems
Proceedings of the Joint Workshop on Feedback and Synthesis of Linear and Nonlinear Systems
Bielefeld/Rom
XIII, 284 pages. 1982

Vol. 40: Y.S. Hung, A.G.J. MacFarlane
Multivariable Feedback:
A Quasi-Classical Approach
X, 182 pages. 1982

Vol. 41: M. Gössel
Nonlinear Time-Discrete Systems –
A General Approach by Nonlinear Superposition
VIII, 112 pages. 1982

Vol. 42: Advances in Filtering and Optimal Stochastic Control
Proceedings of the IFIP-WG 7/1 Working Conference
Cocoyoc, Mexico, February 1–6, 1982
VIII, 391 pages. 1982

For information about Vols. 1–21 please contact your bookseller or Springer-Verlag.

Lecture Notes in Control and Information Sciences

Edited by M. Thoma

74

K. Ichikawa

Control System Design based on Exact Model Matching Techniques

Springer-Verlag Berlin Heidelberg GmbH

Author
Kunihiko Ichikawa
Dept. of Mechanical Engineering
Sophia University
Chiyoda-ku, Tokyo
Japan

DOI 10.1007/978-3-540-39649-9

Library of Congress Cataloging in Publication Data

Lecture Notes in Control and Inf. Sciences 74
Ichikawa, Kunihiko
Control system design based on exact model matching techniques.
(Lecture notes in control and information sciences ; 74)
1. Automatic control. 2. Adaptive control systems.
3. Discrete-time systems. 4. System design.
I. Title. II. Series.
TJ233.I27 1985 629.8 85-17242

Originally published by Springer-Verlag Berlin Heidelberg New York in 1985.
MyCopy version of the original edition 1985
2161/3020-543210
www.springer.com/mycopy

PREFACE

Control system design based on exact model matching techniques is a systematic and comprehensive design techniques with unique and orignal concepts. The exact model matching is a steadyfast design method for automatic control systems unlike the conventional lead, lag, or feedback compensation methods. Of course, it is quite different from trial and error methods. Not only the exact model matching is valuable in itself, but it is a starting point for almost all design methods of higher class such as adaptive control, decoupling control, discrete time system, stabilizing control of unstable delay systems and so on.

Any advanced mathematics are used in this text book. The first chapter concerns with the concept of control system design, focusing on the determination of the reference model. The second chapter demonstrates the state space method for pole assignment and exact model matching. The third chapter is the basis and core of the text book. The frequency domain exact model matching techniques are explained here clearly. The techniques are immediately extended to adaptive control in the fourth chapter. Many adaptive control papers have been published hitherto, but almost all of them are complicated and hard reading, belonging to the same party. In this chapter, adaptive control is demonstrated from the view point that adaptive control is a natural extension of the exact model matching. The exact model matching techniques are further extended to the decoupling control of multivariable systems in the fifth chapter. Almost all papers on decoupling control in the past employed state monstrated in the sixth chapter. The exact model matching techniques are further extended to the decoupling control of multivariable systems in the seventh chapter. Almost all papers in the past on decoupling control employed space techniques without any effective design method except the necessary and sufficient conditions for decoupling by using only state variable feedback. By using exact model matching techniques, the vast range of decoupling problems such that inputs are more than outputs and or matrix denoted by B^* is singular are solved quite easily. Adaptive control for multivariable systems is presented in eighth chapter as the form of natural extension of decoupling control as well as scalar adaptive easily. In the ninth chapter, discrete time systems are dealt with. The famous finite time settling problem is demonstrated as a trivial problem, because the problem is nothing but a choice of the reference model in the exact model matching techniques. The tenth chapter deals with delay systems. The classical Smith's method is confined to the stable plant, and is not considered a general theory. We bear unstable delay plant in mind in this chapter. Recently stabilizing control of unstable delay plant has been established using the state space techion of the exact model matching. The frequency domain exact model matching techniques can solve not only finite pole assignment but also exact model matching for delay system with ease. Furthermore, the techniques are also extended to the adaptive control of delay systems.

This text book is quite peculiar to all other text books ever published. It is written under quite different philosophy on control system design from usual text books. The peculiarity is not in the simplicity of the mathematical tool employed, but is in the design philosophy itself. The system of control system design method presentedin this text book is the fruit of endeavour and deliberation of the author extending over seven years, and is backed up by the teaching experience of twenty-five year in both Nagoya University and Sophia University.

K. Ichikawa

CONTENTS

Chap.1 Introduction 1

Chap.2 Time domain exact model matching 10

2.1 Pole assignment 11

2.2 Exact model matching 15

Chap.3 Frequency domain exact model matching 20

3.1 Pole assignment 20

3.2 Exact model matching 22

Chap.4 Adaptive control 34

4.1 Adaptive identification theory 35

4.2 Adaptive control 43

Chap.5 Disturbance in exact model matching and control 53

5.1 The effect of disturbance 53

5.2 Suppression of disturbance effect 55

5.3 Representation of disturbance and its generation 56

5.4 Control law for the specified disturbance 57

5.5 Suppression of disturbance effect in adaptive control 59

Chap.6 Adaptive pole assignment 62

6.1 Problem statement 62

6.2 Adaptive pole assignment 63

Chap.7 Decoupling control with exact model matching 70

7.1 Description of the problem 71

7.2 Dynamic compensation 76

7.3 Solution of P1 problem 77

7.4 Solution of P2 problem 81

8.5 Generic multivariable exact model matching 84

Chap.8 Multivariable adaptive control 87

8.1 Control law 87

8.2 Error dynamics ········ 89
8.3 Adaptive law ········ 91
Chap.9 Discrete time system ········ 94
9.1 Introduction ········ 94
9.2 Pole assignment ········ 101
9.3 Exact model matching ········ 103
9.4 Design of finite time settling system ········ 104
9.5 Adaptive control ········ 106
Chap.10 Time delay system ········ 117
10.1 Time delay systm and its description ········ 117
10.2 Stabilizing control of delay systems in the time domain ········ 117
10.3 Poles of delay system ········ 119
10.4 Frequency domain finite pole assignment ········ 119
10.5 Frequency domain exact model matching ········ 122
10.6 Adaptive control ········ 125

Chapter 1 Introduction

In the design of a control system, it is required to design a controller such that requirements for the closed loop system characteristics are satisfied, given the controlled object or plant. There exist several kinds of plant such as scalar or multivariable systems as well as continuous or discrete time systems. Also, there exist finite dimensional systems as well as infinite dimensional parameter systems such as delay or distributed parameter systems. Linear time invariant continuous time scalar systems are first considered.

The control system design procedure cannot be started unless the characteristics of the resulting products, the closed loop system in this case, are not specified, as usual machine designs. Although the requirements for the products, the specifications, may involve volume, weight, kind of power to be used and durability, only the closed loop system characteristies from the view poimt of control engineering are considered here. The characteristics are put in order as follows:

(1) speed of response

(2) stabilty or damping property

(3) steady state property

The orderers who are not always well acquaited with control engineering may state their own demands in various forms. The control system designer, however, must interprete those demands into each item of the above three major properties.

The design method based on the classical control theory is a trial and error method. For example, in the first a feedback control system is constructed by introducing a haphazard gain. Since the plant is known, the closed loop system characteristics can be analyzed by either graphical analysis or computer simulation. The resulting characteristics are then compared with the given specifications. If the specifications are not met, the gain will be altered or phase lead and/or phase lag compensation scheme is adopted when the mere gain alteration does not seem to be satisfactory. The closed loop system characteristics are again analyzed and compared with the specifications, and the gain as well as time constants in the phase lead or lag networks are revised. These analyses and alterations of the parameters in the tentative controller are repeated until the specifications are met. Employment of PID-controller is based on quite the same ideas. Can these methods always succeed in leading a controller such that the specifications are met? Further, is there any sure rule for parameter alteration in order to lead a controller? The answers are negative in geneal.

The basis of the method explained in this text is the determination of the desired transfer function (called reference model in the field of adaptive control) and the design procedure is called exact model matching. The designer should not start the design work from the given specifications without preparation, but make a reference model, the transfer function of which satisfies the given specifications well; that is, the desired closed loop system (plant + controller) transfer function. Then, a controller which makes the closed loop system transfer function be identical with reference model is determined according to exact model matching. The determination of a controller means the determination of a control law. The structure of the control system is generally as shown in Fig.1.1 with somewhat modification as the case may be.

Speed of response and stability are usually specified as distinctive features on the step response wave form of Fig.1.2 or the frequency response wave form of Fig.1.3. The overshoot PO on the step response wave form represents the stability of the system purely. The delay time t_d, rise time t_r, and time to peak t_p all represent speed of response almost purely. The settling time t_s depends on both speed of response and stability. On the other hand, M_p appearing in the frequency response wave form represents stability purely, while peak frequency ω_p and bandwidth BW represent speed of response. The steady state property does not appear in these wave forms.

Stabilty property is specified to some value which is desired to be met precisely . On the other hand, speed of response is specified by an inequality such that, for example, t_d should be less than a certain number. However, the fast speed of response does not always lead to a preferable servomechanism. It is undesirable for the servomechanism to respond to noise sensitively because of having too fast speed of response. That is, the servomechanism should have low pass filter property in order to block high frequency noise. Thus, the speed of responce is specified such that, for example, t_d should be less than a certain number while BW should be less than another certain number.

The purpose to specify the steady state property is

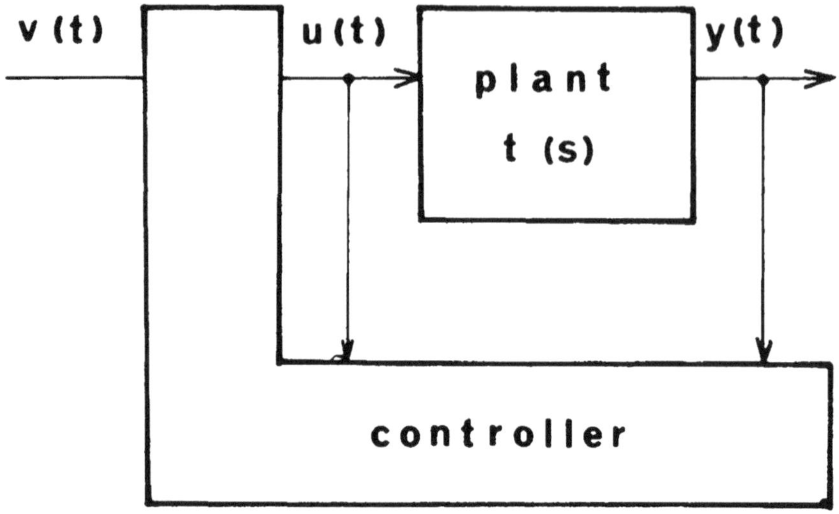

Fig. 1.1 Structure of control system

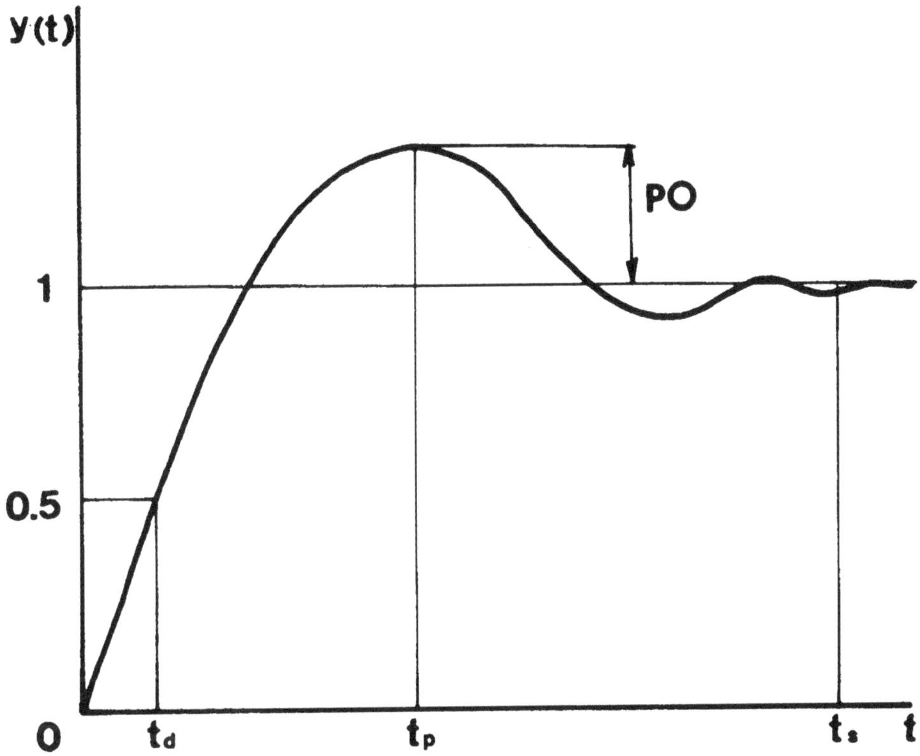

Fig. 1.2 Step response wave form

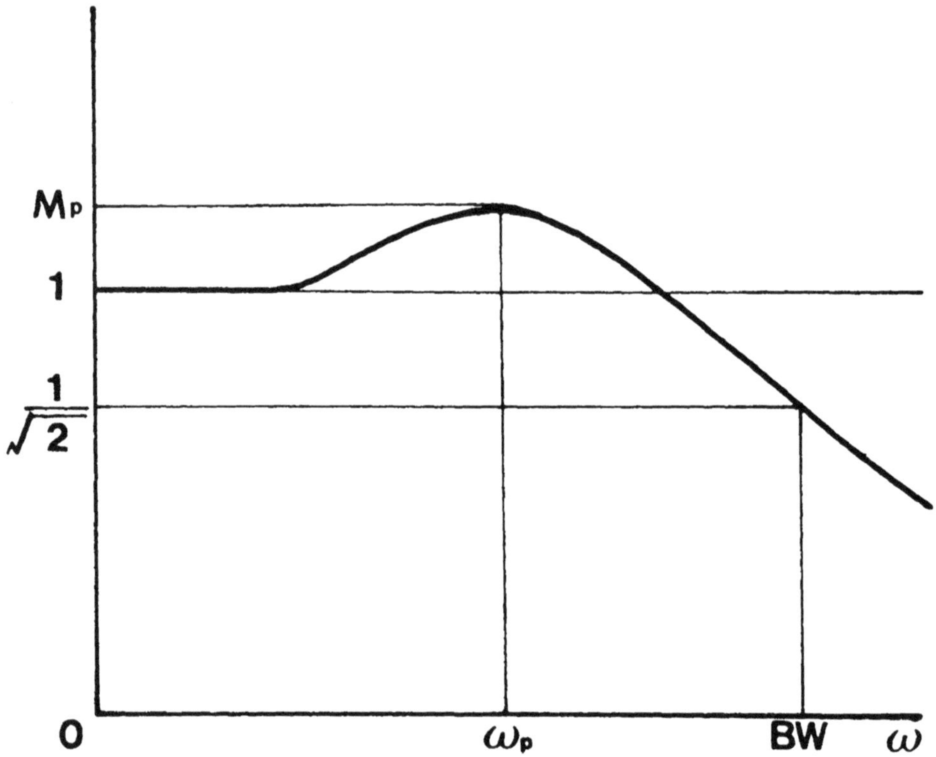

Fig. 1.3 Frequency response wave form

to control steady state ($t\to\infty$) errors resulting from step input, ramp input and so on. For example, the servo is required to have null offset. The ramp input will seldom be applied to the servo. Why, then, does the steady state error due to ramp input be brought into question ? The reason is that the servo is required to be stiff; that is, robust to disturbance with constant value.

Let the closed loop system transfer function be $t_d(s)$, and denote the reference input and plant output by $v(s)$ and $y(s)$ respectively. Then, $e(s)/v(s)=1-t_d(s)$ is obtained, where $e(s)$ is an error. By reducing it to a common denominator and performing a division algorithm after rearranging both numerator and denominator in the ascent order of powers of s, we obtain

$$e(s)/v(s) = \alpha_0+\alpha_1 s+\alpha_2 s^2+\cdots \qquad (1.1)$$

Error constants K_p, K_v, K_a $\cdots$ are defined by the following equations:

$$1/(1+K_p)=\alpha_0,\ 1/K_v=\alpha_1,\ 1/K_a=\alpha_2,\cdots \qquad (1.2)$$

These constants are called position error constant, velocity error constant, acceleration error constant and so on, respectively [1]. These constants were formerly defined from $t(s)$, the open loop transfer function, such as

$$K_p=\lim_{s\to 0} t(s),\ K_v=\lim_{s\to 0} st(s),\ K_a=\lim_{s\to 0} s^2t(s),\cdots \qquad (1.3)$$

[2].

Now, if all error constants were specified, $t_d(s)$ would be fixed, and additional specification for speed of response and/or stability become meaningless. It is usual, however, to specify only K_p and K_v, or at most one more constnt K_a as for steady state property.

The steady state error due to step input $v(s)=1/s$ is calculated as

$$\lim_{t\to\infty} e(t) = \lim_{s\to 0} s(\alpha_0+\alpha_1+\cdots)/s \qquad (1.4)$$

Therefore the condition of null offset requires $K_p=\infty$. Since $\alpha_0=1-t_d(0)$, the condition requires $t_d(0)=1$; i.e., the constant terms in both numerator and denominator must be equal.

The steady state error due to ramp input $v(s)=1/s^2$ is α_1 or $1/K_v$ when $\alpha_0=0$. As mentioned before, in order for the servo to be somehow stiff, α_1 must be less than a certain number, or K_v must be greater than a certain number. By calculating from the relation $\alpha_1 = \{[1-t_d(s)]/s\}_{s=0}$, we obtain

$$\frac{1}{K_v}=\sum_{i=1}^{n} -(\text{inverse of pole})-\sum_{i=1}^{m} -(\text{inverse of zero}) \qquad (1.5)$$

[1],[3]. That is, the specification of the velocity error constant over a certain number means to impose an inequality among the poles and zeros of $t_d(s)$. Moreover, we can obtain the relation

$$\frac{1}{K_a}=\sum_{i=1}^{n} (\text{inverse of pole})^2-\sum_{i=1}^{m} (\text{inverse of zero})^2 + \frac{1}{K_v^{\,2}} \quad . \qquad (1.6)$$

Therefore, thespecification of $|K_a|$ over a certain number will impose another inequality among the poles and zeros of $t_d(s)$. Summarizing above, the following four items are specified in the typical design specifications.

(1) stability: PO = a certain number

(2) speed of response (1): t_d < a certain number

(3) speed of response (2): BW < a certain number

(4) steady state property: $K_p=\infty$, K_v > a certain number

The standard form of the reference model to be determined is as follows:

$$t_d(s)=\frac{\omega_n^2(-\lambda_k)}{-\sigma}\cdot\frac{s-\sigma}{(s^2+2\zeta\omega_n+\omega_n^2)(s-\lambda_k)} \tag{1.7}$$

[4]. In [4] graphs are avilable which aid to determine four parameters ζ, ω_n, λ_k, and σ.

REFERENCES

[1] J.G.Truxual; Control system synthesis p.80, McGraw-Hill, 1955.

[2] G.S.Brown and D.P.Campbell; Principle of servomechanims p.167, John Wiley & Sons, 1948.

[3] K.Ichikawa; The newest control theory p.119 (in Japanese), Gakkensha, 1983.

[4] J.L.Melsa and D.G.Schultz; Linear control systems Chap.8.5, McGraw-Hill, 1969.

Chapter 2 Time domain exact model matching

Let the plant motion be described in state space representation, or by a transfer function $t(s)=gr(s)p^{-1}(s)$. Also, let the reference model be given by $t_d(s)=g_d r_d(s) \times p_d^{-1}(s)$. To determine the controller so that the closed loop transfer function coincides exactly with $t_d(s)$ is called exact model matching. This is the same as to say that the plant input u(t) should be synthesized so that the plant output y(t) tracks asmptotically the reference model output $y_m(t)$. No simpler and plainer principle for control system design than this exists, provided that $t_d(s)$ is determined such that the specifications are satisfied as mentioned in Chapter 1.

There had been endeavors to execute control system design from the above mentioed viewpoint before the advent of modern control theory [1], but they did not achieved a success. The reason is that this seemingly simple problem can never be resolved unless the concept of the state is grasped clearly. Exact model matching is, therefore, is originally resolved by state space method (also called time domain method). Subsequently, the more convenient solution method in s domain was developed, which is called frequency domain method. The credit for the development of frequency domain methods belongs almost to Wolovich [2]. The frequency domain method is superior to the state space method, but can be understood only after the time domain methods are understood. Both methods look like rather different to each other superficially, but are the same in essence. The frequency domain method, however, is simpler in the computational work, and, moreover, plays a starting point of the extension to the more sophisticated control such as adaptive control and decoupling control.

In both methods, the fundamental concept is the linear state variable feedback. Although exact model matching cannot be achieved by state feedback only, it is related closely to state feedback, and is achieved by {precompensator + state feedback} for the augmented plant after providing a suitable precompensator.

2.1 Pole assignment. For the time being, the state $x(t)$ is assumed to be available. Let the plant transfer function be given by $t(s)=gr(s)p^{-1}(s)$, where both $p(s)$ and $r(s)$ are monic with $\partial[p(s)]=n$ and $\partial[r(s)]=m$. Let the state space representation correspoding to $t(s)$ be

$$\dot{x}(t) = Ax(t) + bu(t) \tag{2.1a}$$

$$y(t) = c^T x(t), \tag{2.1b}$$

where $\{c^T, A, b\}$ is controllable and observable by the assumption that $r(s)$ and $p(s)$ are relatively prime. To place all poles of the closed loop system on each preassigned location is called arbitrary pole assignment.

In the following, it will be explained that the necessary and sufficient condition for arbitrary pole assignment is the controllability of (A, b). Suppose that the state feedback

$$u(t) = f^T x(t) + v(t) \tag{2.2}$$

is applied to the plant (2.1), where $v(t)$ is a reference input. The motion of the closed loop system is described by

$$\dot{x}(t) = (A+bf^T)x(t) + bv(t) \tag{2.3a}$$

$$y(t) = c^T x(t). \tag{2.3b}$$

Denote the closed loop transfer function by $t_f(s) = g_f r_f(s) p_f^{-1}(s)$, then

$$\begin{aligned} p_f(s) &= |sI-A-bf^T| \\ g_f r_f(s) &= c^T \operatorname{adj}(sI-A-bf^T)\, b. \end{aligned} \tag{2.4}$$

In the first, let us examine $p_f(s)$.

$$\begin{aligned} p_f(s) &= |sI-A| \cdot |I-(sI-A)^{-1} bf^T| \\ &= p(s) - f^T \operatorname{adj}(sI-A)\, b. \end{aligned} \tag{2.5}$$

The second term of $p_f(s)$ is of the same form as $c^T \times \operatorname{adj}(sI-a)\, b$, the numerator of the plant transfer function, it must be of degree n-1. Hence, $p_f(s)$ is an n degree monic plynomial likewise p(s). The pole assignment is nothing but to specify $p_f(s)$ to some n degree monic polynomial. Since p(s) is already given, it is equivalent to specify $p(s)-p_f(s)=f^T \operatorname{adj}(sI-A)\, b$ such that

$$f^T \operatorname{adj}(sI-A)\, b = \omega_{n-1} s^{n-1} + \cdots + \omega_1 s + \omega_0. \tag{2.6}$$

The vector f which satisfies (2.6) is the required f for state feedback. Since

$$\begin{aligned} f^T \operatorname{adj}(sI-A)\, b &= f^T b s^{n-1} + f^T (A+a_{n-1} I) b s^{n-2} + \cdots \\ &\quad + f^T (A^{n-1} + a_{n-1} A^{n-2} + \cdots a_1 I) b, \end{aligned} \tag{2.7}$$

we obtain

$$\begin{bmatrix} b^T \\ b^T(A^T+a_{n-1}I) \\ \vdots \\ b^T((A^T)^{n-1}+a_{n-1}(A^T)^{n-2}+\cdots+a_1 I) \end{bmatrix} f = \begin{bmatrix} w_{n-1} \\ w_{n-2} \\ \vdots \\ w_0 \end{bmatrix}. \tag{2.8}$$

Now, since

$$\rho[b,(A+a_{n-1}I)b,\cdots,(A^{n-1}+a_{n-1}A^{n-2}+\cdots+a_1 I)b]$$
$$= \rho[b\ Ab\ \cdots\ A^{n-1}b] = n,$$

the matrix in the left hand side of (2.8) is nonsigular when $\{A,b\}$ is controllable, and hence f can be determined; i.e., arbitrary pole assignment can be achieved. By the way, since the relation

$$[(A^{n-1}+a_{n-1}A^{n-2}+\cdots a_1 I)b,\cdots,(A+a_{n-1}I)b,b]=Q^{-1} \tag{2.9}$$

holds [3], where Q a transformation matrix into controllable canonical form [2], it results that f may also be determined from the equation

$$f^T = [w_0\ w_1\ \cdots\ w_{n-1}]Q. \tag{2.10}$$

Let us examine how the numerator will vary under the above state feedback. We get

$$g_f r(s) = c^T \operatorname{adj}(sI-A-bf^T)\, b$$
$$= c^T\{Is^{n-1}+[(A+bf^T)+(a_{n-1}-w_{n-1})I]s^{n-2}+\cdots$$
$$+[(A+bf^T)^{n-1}+(a_{n-1}-w_{n-1})(A+bf^T)^{n-2}+\cdots+(a_1-w_1)I]\}b, \tag{2.11}$$

but the term in the braces reduces to $Is^{n-1}+(A+a_{n-1}I)s^{n-2}+\cdots+(A^{n-1}+a_{n-1}A^{n-2}+\cdots a_1 I)$ by using (2.8). Therefore, it results that

$$g_f r_f(s) = c^T \mathrm{adj}(sI-A)b = gr(s), \tag{2.12}$$

i.e., the numerator polynomial remains unaltered. We could also prove that the controllability of $\{A,b\}$ is necessary for arbitrary pole assignment. Thus, we obtain

Theorem. 2.1 (arbitrary pole assignment) The necessary and sufficient condition for arbitrary pole assignment is that $\{A,b\}$ is controllable.

Confining the argument only to find feedback gain vector f, we can present a simpler explanation by using controllable canonical form. The canonical form of the plant (2.1) is

$$\dot{\hat{x}}(t)=\begin{bmatrix} 0 & 1 & 0 & & 0 \\ 0 & 0 & 1 & & 0 \\ \vdots & \vdots & \vdots & \cdots & \vdots \\ 0 & 0 & 0 & & 1 \\ -a_0+\hat{f}_1 & \cdots & & & -a_{n-1}+\hat{f}_n \end{bmatrix}\hat{x}(t)+\begin{bmatrix} 0 \\ \vdots \\ 0 \\ 1 \end{bmatrix}v(t) \tag{2.13a}$$

$$y(t)=[\hat{c}_0 \quad \hat{c}_1 \quad \cdots \quad \hat{c}_{n-1}]\hat{x}(t). \tag{2.13b}$$

From this, it is easy to see that the numerator polynomial of the closed loop system transfer function is the same as that of plant transfer function, and moreover the denominator polynomial is calculated with ease; i.e.,

$$t_f(s) = \frac{gr(s)}{s^n+(a_{n-1}-\hat{f}_n)s^{n-1}+ \cdots +(a_0-\hat{f}_1)} . \tag{2.14}$$

Thus, the feedback gain $f=[f_1 \; f_2 \cdots f_n]$ to achieve the desired pole assignment is determined as

$$\hat{f} = [w_0 \; w_1 \cdots w_{n-1}]. \tag{2.15}$$

The feedback gain for the original state $x(t)$ is determined by $f^TQ^{-1}=\hat{f}^T$ because $\hat{x}(t)=Qx(t)$. This results agrees with (2.10).

2.2 Exact model matching. Let the plant transfer function be $gr(s)p^{-1}(s)$. The pole assignment scheme using state feedback cannot alter the numerator polynomial, and moreover the denominator polynomial remains as n degree monic polynomial like the plant transfer function. In exact model matching, the reference model is given as

$$t_d(s) = g_d r_d(s) p_d^{-1}(s), \tag{2.16}$$

where both $r_d(s)$ and $p_d(s)$ are monic with $\partial[r_d(s)]=m_d$ and $\partial[p_d(s)]=n_d$. Clearly, the mere state feedback cannot achieve exact model matching. In the following, it is shown that an appropriate pole assignment for the augmented plant which is formed by connecting serially a suitable precompensator to the plant will achieve exact model matching. Two assumptions are made here.

A.1 $r(s)$ has no zeros in the open right half plane.
A.2 $n_d - m_d \geq n - m$.

The assumption A.1 can be rephrased that the plant is of nonminimal phase. We introduce a precompensator with the transfer function $t_c(s) = r_d(s)p_c^{-1}(s)$, where $p_c(s)$ is any monic polynomial of degree $n_d-(n-m)$. The arbi--trariness of $p_c(s)$ comes from the subsequent pole assignmnt. The properness and hence the realizability of $t_c(s)$ is assured, since $\partial[p_c(s)]-\partial[r_d(s)] \geq 0$. The augmentd plant transfer function is

$$t(s)t_c(s) = \frac{gr(s)\; r_d(s)}{p(s)\; p_c(s)} \tag{2.17}$$

with the order of n_d+m. By the arbitrary pole assignment technique mentioned in the previous section, the numerator polynomial can be altered to $r(s)p_d(s)$. Also, it is easy to change the gain from g to g_d. Thus, the closed loop system transfer function will be

$$t_f(s) = \frac{g_d r(s)\; r_d(s)}{r(s)\; p_d(s)}. \tag{2.18}$$

Now, since r(s) is stable by A.1, the closed loop transfer function reduces to $t_d(s)$ after pole zero cancellation; i.e., exact model matching is achieved.

In the time domain method, the above work is executed using state space representation. The plant state x(t) is assumed to be available for the time being. Let the motion of the plant is expressed by

$$\dot{x}(t) = Ax(t) + bu(t) \tag{2.19a}$$

$$y(t) = c^T x(t), \tag{2.19b}$$

where $\{c^T, A, b\}$ is assumed to be controllable and observable. Let the state space representation of the precompensator be

$$\dot{x}_c(t) = A_c x_c(t) + b_c \bar{u}(t) \tag{2.20a}$$

$$u(t) = c_c^T x_c(t) + d_c \bar{u}(t). \tag{2.20b}$$

The output of the precompensator is used as the plant input, and hence the output must be u(t). Also, $\bar{u}(t)$, the input to the precompensator, is an input to the augmented plant. It is to be noticed that the precompensator is not always strictly proper. The state space representation of the augmented plant becomes

$$\begin{bmatrix} x(t) \\ x_c(t) \end{bmatrix} = \begin{bmatrix} A & bc^T \\ 0 & A_c \end{bmatrix} \begin{bmatrix} x(t) \\ x_c(t) \end{bmatrix} + \begin{bmatrix} d_c b \\ b_c \end{bmatrix} \bar{u}(t) \tag{2.21a}$$

$$y(t) = [c^t \quad 0] \begin{bmatrix} x(t) \\ x_c(t) \end{bmatrix}. \tag{2.21b}$$

By the routine for pole assignment, it is easy to find a feedback gain vector $[f^T \ f_c^T]^T$ so that the closed loop system characteristic polynomial becomes $r(s)p_d(s)$.

In reality, x(t) is unavailable, and hence it is necesary to provide an state observer to use the estimate $\tilde{x}(t)$; on the other hand, since the precompensator is constructed artificially, $x_c(t)$ is available. Thus, the control law based on state feedback will be as follows;

$$\bar{u}(t) = [f^T \quad f_c^T] \begin{bmatrix} \tilde{x}(t) \\ x_c(t) \end{bmatrix} + (g_d/g)v(t). \tag{2.22}$$

Alternatively, $x_a(t)=[x(t) \ x_c(t)]^T$, the state of the augmented plant, may be estimated collectly to obtain $\tilde{x}_a(t)$. In this case, the control law becomes

$$\bar{u}(t) = f_a^T\tilde{x}_a(t) + (g_d/g)v(t). \tag{2.23}$$

Both exact model matching systems are shown in Figs. 2.1 and 2.2 respectively.

REFERENCES

[1] J.G.Truxal;Control system synthesis, Chap.5, McGraw-Hill,1955.

[2] W.A. Wolovich;Linear multivariable systems, Springer-Verlag,1974.

[3] K. Ichikawa;Introduction to control engineering, Sangyotosho,1978.

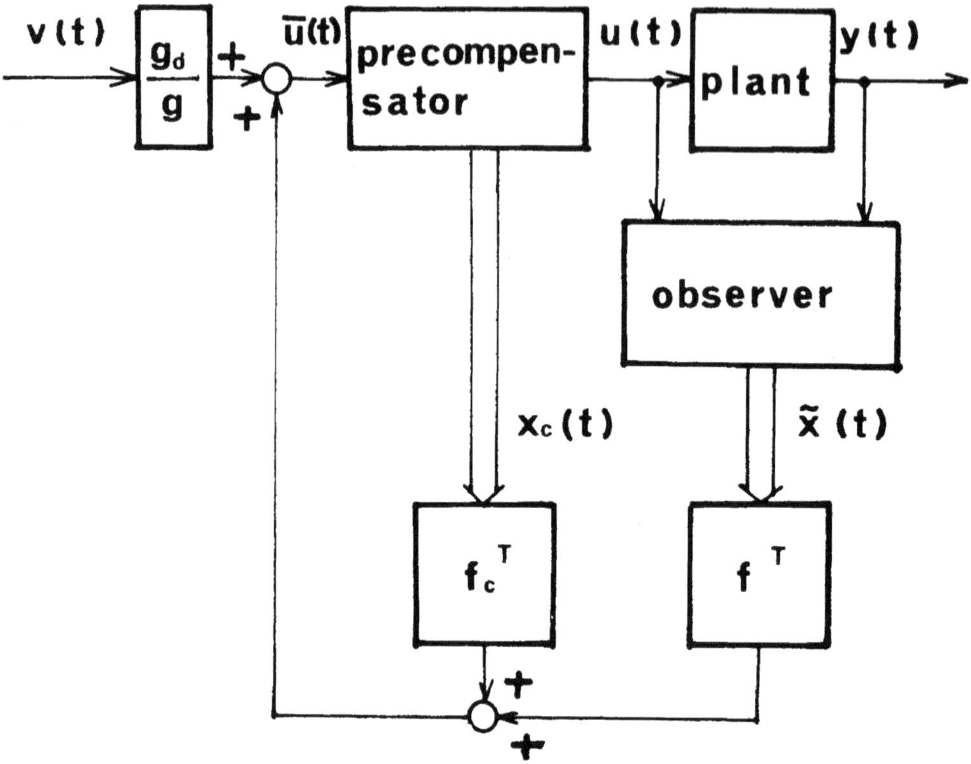

Fig. 2.1 Exact model matching with an observer for x(t)

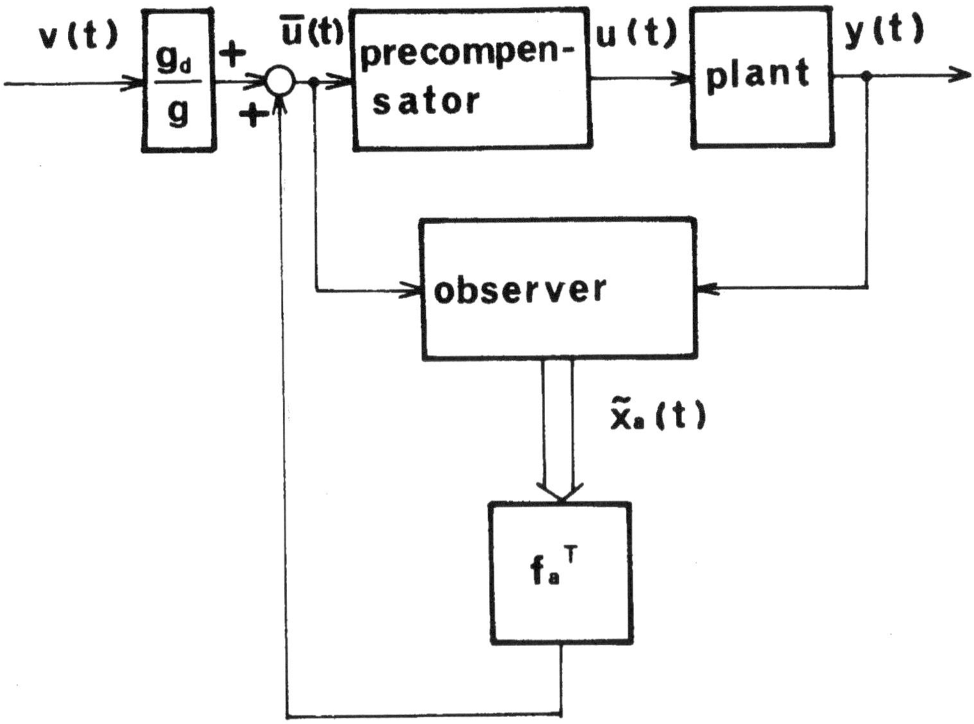

Fig. 2.2 Exact model matching system with an observer for $[x^T(t)\ x_c^T(t)]^T$

Chapter 3 Frequency domain exact model matching

3.1 Pole assignment. Let the plant transfer function be $t(s)=gr(s)p(s)$. Denote an arbitrary n degree monic polynomial by $p_f(s)$. The pole assignment is meant by producing a closed loop system with transfer function of $g_f r(s) \times p_f^{-1}(s)$. It is to be noticed that $f(s)=p(s)-p_f(s)$ is of n-1 degree.

Lemma 3.1. [1] (Existence of polynomials q(s),k(s), and h(s)) Assume that gr(s) and p(s) are relatively prime, there exists a triple of polynomials {q(s),k(s),h(s)} satisfying the following three conditions for any n-1 degree poly-nomial f(s):

(1) q(s) is aymptotically stable.

(2) $k(s)p(s) + h(s)gr(s) = q(s)f(s)$ (3.1)

(3) Both $q^{-1}(s)k(s)$ and $q^{-1}(s)h(s)$ are proper.

(proof) By taking q(s) as any n-1 degree stable polynomial, (1) is satisfied. Since q(s)f(s) is of degree 2n-2, it can be represented uniquely as the sum $\alpha_1(s)gr(s)+\alpha_2 p(s)$ with both $\alpha_1(s)$ and $\alpha_2(s)$ are of n-1 degree. By rewriting $\alpha_1(s)$ and $\alpha_2(s)$ as h(s) and k(s) respectively, (2) is assured. Moreover, it is clear that both $q^{-1}(s)k(s)$ and $q^{-1}(s)h(s)$ are proper. Q.E.D.

In reality, since q(s)f(s) are of 2n-2 degree and gr(s) is of degree n-1, k(s) must be of degree n-2, and hence $q^{-1}(s)k(s)$ becomes strictly proper.

Theorem 3.1. [1] (pole assignment) Using $q(s)$, $k(s)$, and $h(s)$ defined in Lemma 3.1, the control law

$$u(s) = \frac{k(s)}{q(s)}u(s) + \frac{h(s)}{q(s)} + \frac{g_f}{g}v(s) \qquad (3.2)$$

is feasible and achieves pole assignment.

(proof) Since both $q^{-1}(s)k(s)$ and $q^{-1}(s)h(s)$ are proper, the control law (3.2) is feasible. From eq.(3.1) we obtain

$$k(s)u(s) + h(s)gr(s)p^{-1}(s)u(s) = q(s)f(s)p^{-1}(s)u(s).$$

However, since $gr(s)p^{-1}(s)u(s)=y(s)$, this becomes

$$k(s)u(s) + h(s)y(s) = q(s)f(s)p^{-1}(s)u(s).$$

On the other hand, we obtain from eq.(3.2)

$$q(s)u(s) = k(s)u(s) + h(s)y(s) + (g_f/g)q(s)v(s).$$

From the above two equations, we obtain

$$q(s)u(s) = q(s)f(s)p^{-1}(s)u(s) + (g_f/g)v(s),$$

or in other words,

$$p^{-1}(s)u(s) = p_f^{-1}(s)q^{-1}(s)q(s)(g_f/g)v(s).$$

Then,

$$y(s) = gr(s)p^{-1}(s)u(s) = g_f r(s)p_f^{-1}(s)q^{-1}(s)q(s)v(s).$$

However, since $q(s)$ is stable, pole zero cancellation is allowed and finally we obtain

$$\frac{y(s)}{v(s)} = \frac{g_f r(s)}{p_f(s)}. \qquad \text{Q.E.D.}$$

Since the plant is of order n and the control law is

implemented by the system of order n-1, the closed loop system is of order 2n-1, which appears in the transfer function before cancelling q(s). The structure of pole assignment system is shown in Fig. 3.1. Comparing it with the pole assignment system constructed in time domain, it can be said that frequency domain pole assignment automatically yields observer of order n-1, the minimal order state observer. The fact that q(s) may be arbitrary as long as it is stable corresponds to the fact that the pole configuration of the observer is arbitrary as long as all poles lie in the left half plane.

3.2 Exact model matching. Since the frequency domain method is essentialy the same as time domain method, the formula "exact model matching = precompesator + pole assignment" is again true. Also the assumptions A.1 and A.2 mentioned in Sec. 2.2 are required again. The precompensator $t_c(s)=r_d(s)p_c^{-1}(s)$ is provided, where $p_c(s)$ is any $n_d-(n-m)$ degree monic polynomial. The combined system $t(s)t_c(s)$ means an augmented plant, whose denominator polynomial $p(s)p_c(s)$ is of degree n_d+m. On performing the pole assignment so that the denominator polynomial be $r(s)p_d(s)$, the closed loop system transfer function becomes

$$\frac{y(s)}{v(s)} = \frac{g_d r(s)\ r_d(s)}{r(s)\ p_d(s)} \tag{3.3}$$

after cancelling q(s). Since r(s) is stable by A.1, this reduces to $t_d(s)$.

Speaking concretely, the design procedure is as follows. Put $f(s)=p(s)p_c(s)-r(s)p_d(s)$, which is of degree n_d+m-1. Choose q(s) as any monic stable polynomial of

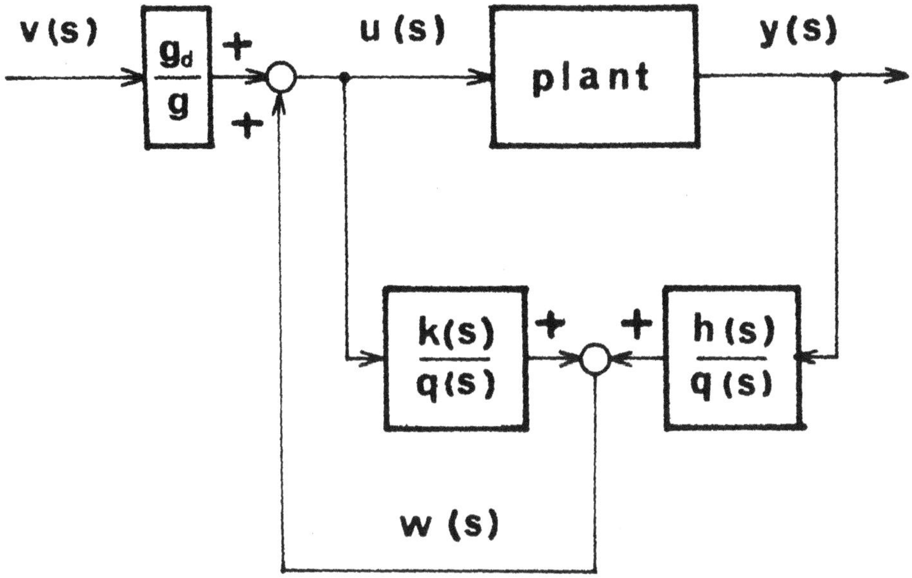

Fig. 3.1 Pole assignment system designed in frequency domain

degree n_d+m-1. The polynomials $k(s)$ and $h(s)$ are determined from the equation

$$k(s)p(s)p_c(s) + h(s)gr(s)r_d(s) = q(s)f(s). \qquad (3.4)$$

Then, $k(s)$ and $h(s)$ are determined as n_d+m-2 and n_d+m-1 degree polynomials respectively. The control law should specify the augmented plant input $\bar{u}(t)$, i.e., the input to the precompensator. Clearly, it must be

$$\bar{u}(s) = \frac{k(s)}{q(s)}\bar{u}(s) + \frac{h(s)}{q(s)}y(s) + \frac{g_d}{g}v(s). \qquad (3.5)$$

The resulting exact model matching system is shown in Fig. 3.2, which corresponds to not Fig. 2.1 but to Fig. 2.2 designed in time domain.

If $r_d(s)$ and $p(s)$ are not relatively prime, the numerator and denominator polynomials of the augmented plant become unprime, and some modification is required for pole assignment. As a simple method, unity feedback around the plant is employed, because then the transfer function of the modified plant becomes $gr(s)[p(s)+gr(s)]^{-1}$, where $r_d(s)$ and $p(s)+gr(s)$ are surely relatively prime.

3.3 The second type exact model matching. An exact model matching with more convenient form than the result in the previous section is developed here. This is hard to obtain through time domain technique [2]. It is natural to assume that the reference model is stable; i.e., $p_d(s)$ is stable. Let $r^*(s)$ and $p^*(s)$ be any stable m and n degree polynomials respectively, and rewrite $t_d(s)$ as follows.

$$t_d(s) = \frac{g_d r^*(s)}{p^*(s)} \cdot \frac{r_d(s)p^*(s)}{r^*(s)p_d(s)} . \qquad (3.4)$$

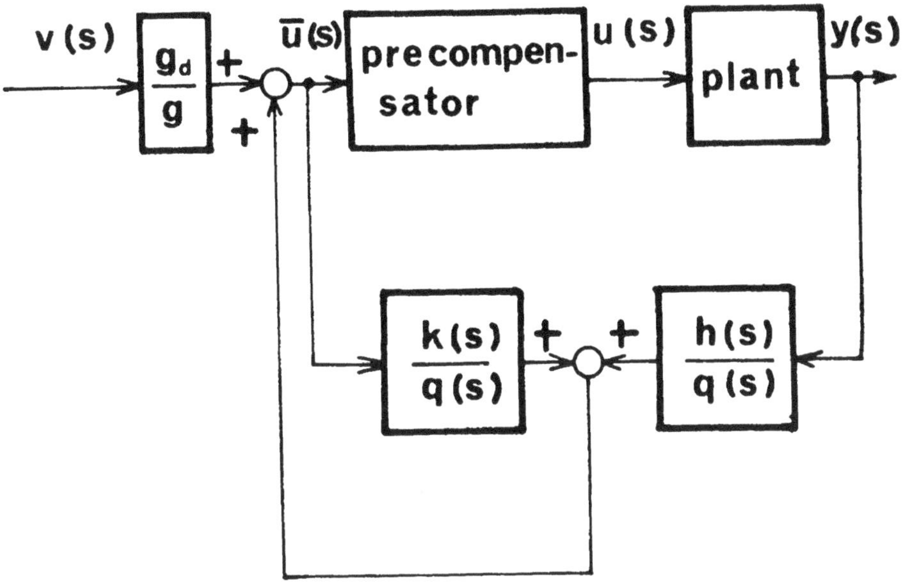

Fig. 3.2 Exact model matching system in frequency domain

Let us examine $t_{IN}(s)=r_d(s)p^*(s)[r^*(s)p_d(s)]^{-1}$. Since $\partial[r^*(s)p_d(s)]-\partial[r_d(s)p^*(s)]=(n_d-m_d)-(n-m)\geq 0$, $t_{IN}(s)$ must be proper and stable, which we call input dynamics. Now, partion $t_d(s)$ as shown in Fig. 3.3, and define $\bar{v}(s)=t_{IN}(s)\times v(s)$. It is to be noticed that input dynamics can be determined irrespective of plant parameters. It is seen that a new exact model matching problem is formed, in which the reference model is $g_d r^*(s)/p^*(s)\overset{d}{=}t^*(s)$ and the reference input is $\bar{v}(s)$. A special feature of this new exact model matching problem is that $\partial[r(s)]=\partial[r^*(s)]=m$, $\partial[p(s)]=\partial[p^*(s)]=n$, and $r^*(s)$ is stable.

Following the procedure mentioned in the previous section, the precompensator is determined as $r^*(s)p_c^{-1}(s)$, where $p_c(s)$ is any m degree monic polynomial. If $p_c(s)$ is particularly chosen as $r^*(s)$, the augmented plant becomes

$$t(s)t_c(s) = \frac{gr(s)r^*(s)}{p(s)r^*(s)}, \tag{3.5}$$

where the numerator and denominator are no longer relatively prime. The exact model matching, however, can be achieved as explained in the following for the reason that the greatest common divisor $r^*(s)$ is stable. In the first, let $\tau_0(s)$ be some n+m-1 degree monic stable polynomial, and determine $k_0(s)$ and $h_0(s)$ from

$$k_0(s)p(s)r^*(s) + h_0(s)gr(s)r^*(s) = \tau_0(s)[r^*(s)p(s)-r(s)p^*(s)], \tag{3.6}$$

where $k_0(s)$ and $h_0(s)$ are n+m-2 and n+m-1 degree polynomials respectively. In order that the equation holds,

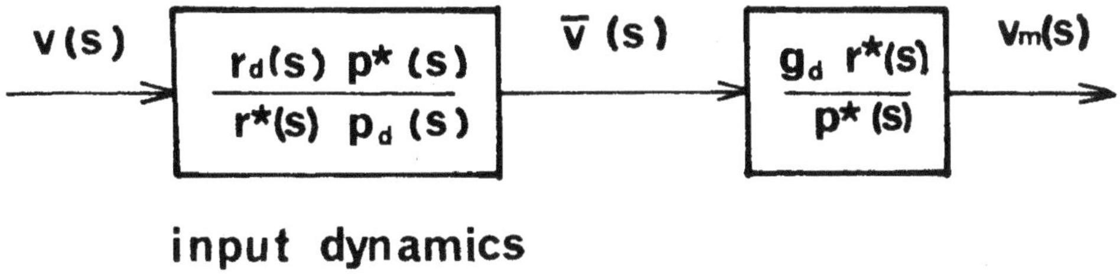

Fig. 3.3 Partition of reference model

$\tau_0(s)$ must be divided by $r^*(s)$. Further, in order to determine $k_0(s)$ and $h_0(s)$ such that both are divisible by $r^*(s)$, $\tau_0(s)$ is assumed to be divisible by $[r^*(s)]^2$. Let

$$\left.\begin{aligned} \tau_0(s) &= \tau(s)[r^*(s)]^2 \\ k_0(s) &= k(s)r^*(s) \\ h_0(s) &= h(s)r^*(s). \end{aligned}\right\} \tag{3.7}$$

Then eq.(3.5) reduces to

$$k(s)p(s)+h(s)gr(s) = \tau(s)r^*(s)p(s)-\tau(s)r(s)p^*(s), \tag{3.8}$$

where $\tau(s)$ is any n-m-1 degree monic stable polynomial. Since the right hand side of eq.(3.8) is of degree 2n-2 and moreover $p(s)$ and $gr(s)$ are relatively prime, n-2 degree $k(s)$ and n-1 degree $h(s)$ are dtermined uniquely, and hence $k_0(s)$ and $h_0(s)$ are determined. Writing down the control law according to the rule mentioned in the previous section, we obtain

$$\begin{aligned} \bar{u}(s) &= \frac{k_0(s)}{\tau_0(s)}\bar{u}(s) + \frac{h_0(s)}{\tau_0(s)}y(s) + \frac{g_d}{g}\bar{v}(s) \\ &= \frac{k(s)}{\tau(s)r^*(s)}\bar{u}(s)+\frac{h(s)}{\tau(s)r^*(s)}y(s)+\frac{g_d}{g}\bar{v}(s). \end{aligned} \tag{3.9}$$

Clearly, the precompensator is $r^*(s)/r^*(s)=1$; i.e., the precompensator is not needed, and $\bar{u}(s)$ is nothing but $u(s)$.

Under the above consideration, we summarize the second type exact model matching as a theorem. In order to keep the unification over multivariable control theory to be mentioned in Chap. 7, let us provide a coefficient $1/g$ before the plant, and regard $t(s)\cdot(1/g)=r(s)p^{-1}(s)$ as a new plant. Denote the input to the coefficient $1/g$ by $u_b(s)$.

Theorem 3.2. (second type exact model matching) Let $\tau(s)$ be any n-m-1 degree monic stable polynomial, and determine $k_b(s)$ and $h_b(s)$ by

$$k_b(s)p(s)+h_b(s)r(s)=\tau(s)r^*(s)p(s)-\tau(s)r(s)p^*(s). \quad (3.10)$$

Then, the control law

$$u_b(s) = \frac{k_b(s)}{\tau(s)r^*(s)}u_b(s) + \frac{h_b(s)}{\tau(s)r^*(s)}y(s) + g_d\bar{v}(s) \quad (3.11)$$

is feasible and achieves exact model matching.

(proof) Since $k_b(s)$ and $h_b(s)$ are determined uniquely as polynomials of degree n-2 and n-1 repectively, and $\tau(s)r^*(s)$ is of degree n-1, the control law is feasible. From eq.(3.10) we obtain

$$k_b(s)u_b(s)+h_b(s)y(s)=\tau(s)r^*(s)u_b(s)-\tau(s)p^*(s)y(s).$$

On the other hand, we obtain from eq.(3.11)

$$\tau(s)r^*(s)u_b(s)=k_b(s)u_b(s)+h_b(s)y(s)+g_d\tau(s)r^*(s)\bar{v}(s).$$

The above two equations are combined to yield

$$y(s) = \frac{g_d\tau(s)r^*(s)}{\tau(s)p^*(s)}\bar{v}(s).$$

However, since $\tau(s)$ is stable, this equation reduces to

$$y(s) = \frac{g_d r^*(s)}{p^*(s)}\bar{v}(s) = \frac{g_d r_d(s)}{p_d(s)}v(s). \quad (3.12)$$

Further, we obtain

$$u_b(s) = \frac{g_d r_d(s)p(s)}{p_d(s)r(s)}v(s).$$

Since $r(s)$ is assumed to be stable, $u_b(t)$ is bounded so long as $v(t)$ is bounded. Thus, the theorem is established. Q.E.D.

The control law with regard to $u(s)$ can be written as

$$u(s) = \frac{1}{g}\{\frac{k(s)}{\gamma(s)r^*(s)}u(s)+\frac{h(s)}{\gamma(s)r^*(s)}y(s)+g_d\bar{v}(s)\}, \quad (3.13)$$

where $k(s)=gk_b(s)$ and $h(s)=h_b(s)$. The resulting system is shown in Fig. 3.4. The input dynamics can be reduced to rather low order system, since both stable polynomials $r^*(s)$ and $p^*(s)$ have considerable freedom. It is to be noticed that the input dynamics introduced by Wolovich depend on the plant parameters, and hence is not suitable for extension to adaptive control. The second type exact model matching mentioned here is qualified to be the basis for various kinds of extension such as adaptive control, decoupling control, etc.

(example) Let a servomotor and reference model transfer functions be given by $t(s)=1.31[s(s+5.45)]^{-1}$ and $t_d(s)=2.5(s+1.6)[(s^2+s+1)(s+4)]^{-1}$ respectively. This $t_d(s)$ was chosen so as to satisfy the specifications that PO=20%, t_d=1(sec), $K_p=\infty$, K_v=1.6(rad/sed). The polynomial $r^*(s)$ is determined as 1, and $p^*(s)$ is chosen as s^2+s+1. Then, the input dynamics are determined as $t_{IN}(s)=(s+1.6)/(s+4)$. Choose $\gamma(s)=s+1$. Then,

$$k_b(s)s(s+5.45)+h_b(s)=(s+1)s(s+5.45)-(s+1)(s^2+s+1),$$

from which $k_b(s)=4.45$ and $h_b(s)=-20.80s-1$ are obtained. The resulting control system is shown in Fig. 3.5.

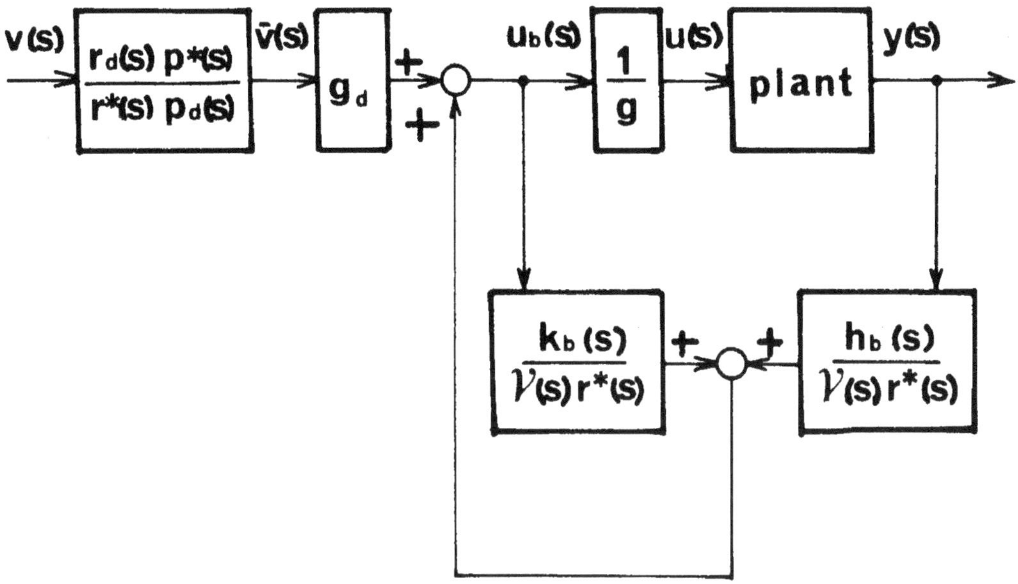

Fig. 3.4 The second type exact model matching system in frequency domain

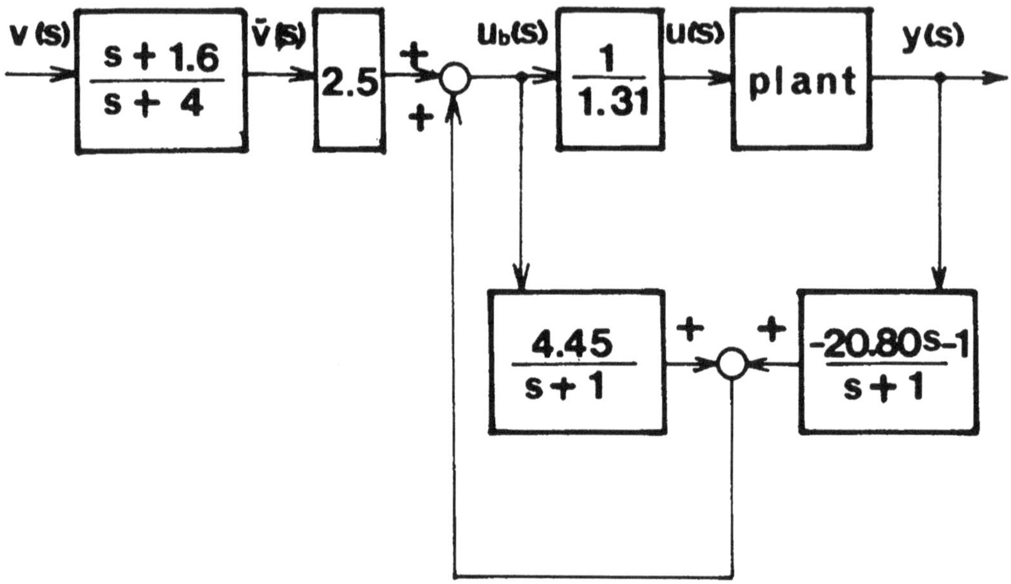

Fig. 3.5 Exact model matching (example)

REFERENCES

[1] W.A. Wolovich;Linear multivariable systems, Spriger-Verlag,1974.

[2] K. Ichikawa; Construction of adaptive control system based on an exact model matching technique,SICE Trans. 20,10,926/931 (in Japanese),1984.

Chapter 4 Adaptive control

The normal way of controlling a plant is to look for the plant transfer function first and then to execute exact model matching. It may be far better if controller parameters could be automatically settled during the operation to the values which would be determined from exact model matching algorithm, after starting the operation of the plant while the plant parameters are unknown. Besides, plant dynamics may vary during its operation owing to the environment conditions. Since exact model matching should be kept continuously to the current plant transfer function in the latter case, exact model matching must be performed automatically again. These control scheme can be regarded as an exact model matching under the condition that the plant transfer function is unknown. Such a control is called adaptive control. In many other literatures, adaptive control is defined as to synthesize plant input u(t) so as to make the plant output follow the reference model output $y_m(t)$ under the condition that the plant transfer function is unknown. That may be true, and exact model matching must not always be achieved in view of the purpose of adaptive control. However, it is evident that for y(t) to follow $y_m(t)$ always for any reference input v(t) means the achievement of exact model matching. By considering adaptive control as mentioned above, adaptive control problem can be grasped on the extension line of exact model matching, which in turn leads to a very plain solution to adaptive control problem. The usual point of view that adaptive control is a problem in the distinct field separates adaptive control from exact model matching theory, and makes it a difficult problem.

The desire to solve adaptive control has been kept since early times. Landau introduced as many as 154 literatures only in the field of adaptive control theory in his survey paper of 1974. However, it can be said that the beginning of adaptive control theory in the strict sense of theory was in 1974, when Monopoli's paper was published. Since then, an immense number of literatures have been published [1]-[13].

In this chapter, a solution to the adaptive control problem is explained as an extension of exact model matching introduced in the previous section, without any reference to the past various theories including self-tuning regulator. That is, the control scheme to be mentioned is a direct method in which adjustable control parameters are adjusted owing to adaptive control error. In this direct scheme, explicit plant identification is not done, but the construction theory of adaptive control system has to be based on identification theory developed for adaptive identification. The adaptive identification theory is stated in Sec. 4.1 and the adaptive control theory is expressed in Sec. 4.2.

4.1 Adaptive identification theory. The first important matter in adaptive identification is how to describe the error dynamics. The second one is to find an effective adaptive law. The adaptive identification theory has been developed for the purpose to identify a plant in real. In the first, the plant dynamics are expressed in the form with seeming statical relation, called non-minimal realization. There have been several methods in this context [14], but the method to be mentioned here is the simplest one. Assume the order of the plant is known, and denote the unknown plant transfer function by

$$t(s) \stackrel{d}{=} \frac{y(s)}{u(s)} = \frac{gr(s)}{p(s)} = \frac{r_{n-1}s^{n-1}+\cdots+r_0}{s^n+p_{n-1}s^{n-1}+\cdots+p_0}, \tag{4.1}$$

which has 2n unknown parameters. Both u(t) and y(t) is available, and u(t) which can be chosen arbitrarily is assumed to be piecewise continuous. Let $\tau(s)$ be any n degree monic stable polynomial. Dividing both numerator and denominator by $\tau(s)$, eq.(4.1) can be rewritten as

$$\frac{y(s)}{u(s)} = \frac{b_{n-1}\frac{s^{n-1}}{\tau(s)}+\cdots+b_1\frac{s}{\tau(s)}+b_0\frac{1}{\tau(s)}}{1-a_{n-1}\frac{s^{n-1}}{\tau(s)}-\cdots-a_1\frac{s}{\tau(s)}-a_0\frac{1}{\tau(s)}}, \tag{4.2}$$

where $b_i=r_i$, $i=1,2,\cdots,n-1$, and $\{a_0,a_1,\cdots,a_{n-1}\}$ corresponds one to one to $\{p_0,p_1,\cdots,p_{n-1}\}$. Then, we obtain

$$y(s) = \sum_{i=0}^{n-1} b_i \frac{s^i}{\tau(s)}u(s) + \sum_{i=0}^{n-1} a_i \frac{s^i}{\tau(s)}y(s). \tag{4.3}$$

We may use Laplace operator s and differential operator p commutatively. Now $p^i/\tau(p)\cdot u(t)$ and $p^i/\tau(p)\cdot y(t)$, $i=1,2,\cdots,n-1$ are obtained by applying u(t) or y(t) to an n-th order filter $1/\tau(s)$ shown in Fig. 4.1. If these signals are defined as those obtained when all initial conditions in integrators are zero, then y(t) when some initial conditions are set in integrators is given by

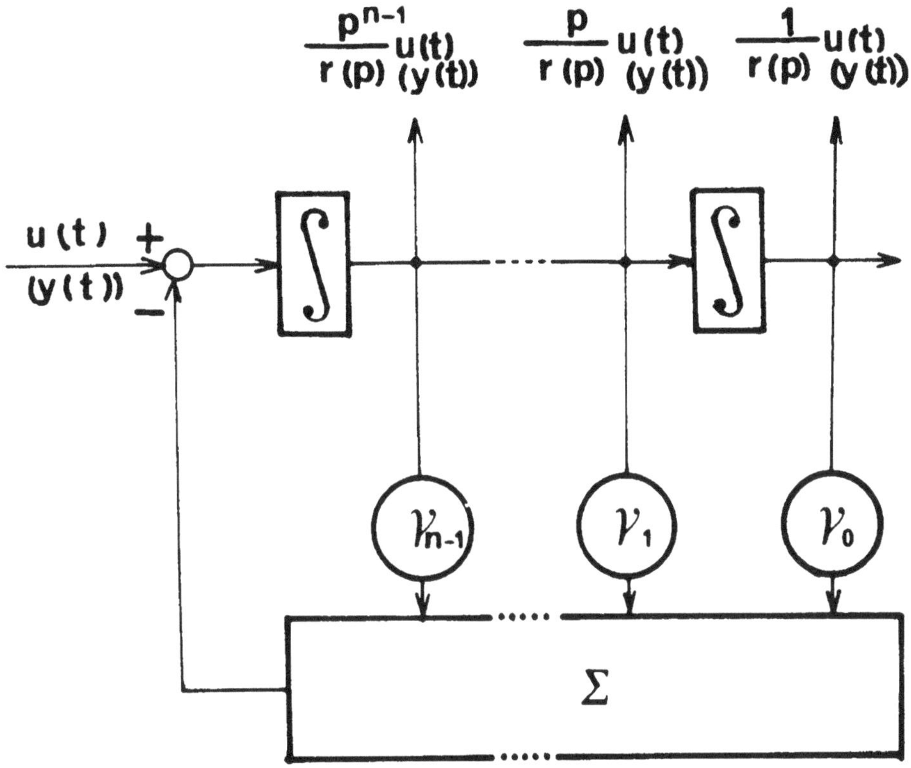

Fig. 4.1 Filter 1/γ(s)

$$y(t) = \sum_{i=0}^{n-1} \{b_i \frac{p^i}{T(p)}u(t) + a_i \frac{p^i}{T(s)}y(t)\} + f(t), \tag{4.4}$$

where f(t) is a time function arising from nonzero initial conditions in integrators. Since $\{y(t),\dot{y}(t),\cdots, y^{(n-1)}(t)\}$, the initial conditions of the plant are unknown, these initial conditions in integrator are also unknown, and hence f(s) is unknown. Also, since $T(s)$ is a stable polynomial, f(t) decays exponentially to zero.

Define

$$\theta = [b_{n-1},\cdots,b_0,a_{n-1},\cdots,a_0]^T \tag{4.5a}$$

and

$$\omega(t) = [\frac{p^{n-1}}{T(p)}u(t),\cdots,\frac{1}{T(s)}u(t),\frac{p^{n-1}}{T(s)}y(t),$$
$$\cdots,\frac{1}{T(s)}y(t)]^T. \tag{4.5b}$$

Then eq.(4.4) can be expressed as

$$y(t) = \theta^T\omega(t) + f(t), \tag{4.6}$$

where $\omega(t)$ is 2n dimensional available signal vector, θ is 2n dimensional unknown parameter vector, and y(t) is detectable as a plant output.

The adaptive identifier is a dynamical system described by

$$\tilde{y}(t) = \tilde{\theta}^T(t)\omega(t) \tag{4.7}$$

which is obtained by replacing the unknown parameter θ, by $\tilde{\theta}(t)$, the adjustble parameter or the estimate, where $\tilde{y}(t)$ can be calculated from eq.(4.7) or measured as an output of a real identifier. The identification error e(t) is defined by

$$e(t) = \tilde{y}(t) - y(t), \tag{4.8}$$

where e(t) is available. Now, put

$$\phi(t) = \tilde{\theta}(t) - \theta, \tag{4.9}$$

where $\phi(t)$ indicates the prameter error vector. Since θ is unknown, $\phi(t)$ is unknown too, but $\dot{\phi}(t)=[\tilde{\theta}(t)]^{\cdot}$ holds. The identification error can be expressed as

$$e(t) = \phi^T(t)\omega(t) - f(t), \tag{4.10}$$

which is the expression of error dynamics to be found.

Since $f(t)$ decays exponential, it will be omitted in the sequel. An adaptive law is a prescription for $\dot{\phi}(t)=[\tilde{\theta}(t)]^{\cdot}$ so as to make $e(t)$ be zero, although the ultimate object is to make $\phi(t)$ tend to zero. Since the available signals are limited to $e(t)$, $\omega(t)$ and $\theta(t)$ itself, the adaptive law should be prescribed in terms of these signals. There are many variations for adaptive laws. The simplest one is

$$\dot{\phi}(t) = -\Gamma\omega(t)e(t), \quad \Gamma=\Gamma^T > 0, \tag{4.11}$$

but it does not provide fast convergence, and moreover it yields some difficulty in proving the boundedness of $\omega(t)$ when it is used in adaptive control.

The adaptive law based on the least square principle yields fast convergence. Omitting $f(t)$, eq.(4.6) becomes $y(t) = \theta^T\omega(t)$, it is necessary to take observation error into consideration in order to derive an adaptive law based on the least square principle. Let us denote the runng time and observation error by τ and $\varepsilon(\tau)$ respectively. The, the plant output is described as

$$y(\tau) = \theta^T\omega(t) + \varepsilon(\tau). \quad 0\leq\tau\leq t \tag{4.12}$$

The least square principle implies that the estimated value of θ should be determined so that $\int_0^t \varepsilon^2(\tau)\, d\tau$ be minimum. By differentiating $\int_0^\tau [y(\tau)-\theta^T\omega(\tau)]^2\, d\tau$ by θ and setting the derivative to 0, the estimate $\tilde{\theta}(t)$ is obtained. That is,

$$\tilde{\theta}(t) = [\int_0^t \omega(\tau)\omega^T(\tau)\, d\tau]^{-1} \int_0^t \omega(\tau)y(\tau)\, d\tau, \tag{4.13}$$

where $\int_0^t \omega(\tau)\omega^T(\tau)\, d\tau$ is assumed to be positive definite. This formula, however, does not assume the form for an adaptive law, and hence we proceed to obtain the form of adaptive law. By differentiating both sides of the relation

$$\int_0^t \omega(\tau)\omega^T(\tau)\, d\tau \cdot \tilde{\theta}(t) = \int_0^t \omega(\tau)y(\tau)\, d\tau \tag{4.14}$$

by t, we obtain

$$\dot{\tilde{\theta}}(t) = -[\int_0^t \omega(\tau)\omega^T(\tau)\, d\tau]^{-1}\{\omega(t)y(t)-\omega(t)\omega^T(t)\tilde{\theta}(t)\} \tag{4.15}$$

However, $\omega^T(t)\tilde{\theta}(t)$ is the identifier output $\tilde{y}(t)$. Then, eq.(4.15) can be rewritten as

$$\dot{\tilde{\theta}}(t) = -[\int_0^t \omega(\tau)\omega^T(\tau)\, d\tau]^{-1}\, \omega(t)\, e(t). \tag{4.16}$$

In order to avoid the operation of inverse and integration, let us introduce a symmetric matrix $\Gamma(t)$ by

$$\Gamma(t) = [\int_0^t \omega(\tau)\omega^T(\tau)\, d\tau]^{-1}. \tag{4.17}$$

Then, eq.(4.16) can be written as

$$\dot{\tilde{\theta}}(t) = -\,\Gamma(t)\,\omega(t)\,e(t). \tag{4.18}$$

On the other hand, we obtain from eq.(4.17)

$$\int_0^t \omega(\tau)\omega^T(\tau)\, d\tau \cdot \Gamma(t) = I. \tag{4.19}$$

Differentiating the above equation by t, we obtain

$$\begin{aligned}\dot{\Gamma}(t) &= -[\int_0^t \omega(\tau)\omega^T(\tau)\, d\tau]^{-1}\, \omega(t)\omega^T(t)\gamma(t)\\ &= -\,\Gamma(t)\omega(t)\omega^T(t)\gamma(t).\end{aligned} \tag{4.20}$$

Since $\lim_{t\to\infty} \int_0^t \omega(\tau)\omega^T(\tau)\, d\tau = 0$, $\Gamma(0)$ should be infinity. In reality, $\Gamma(0)$ is set to αI with α sufficiently large positive number. Thus, the adaptive law based on the least square principle is given by

$$\dot{\tilde{\theta}}(t) = -\Gamma(t)\omega(t)e(t)$$
$$\dot{\Gamma}(t) = -\Gamma(t)\omega(t)\omega^T(t)\Gamma(t) \tag{4.21}$$
$$\Gamma(0) = \alpha I, \quad \alpha >> 1.$$

Theorem 4.1 (adaptive law 1) Assume that $\omega(t)$ is uniformly bounded. Then, adaptive law (4.21) yields $\lim_{t\to\infty} e(t)=0$, $\lim_{t\to\infty} \phi^T(t)\omega(t)=0$, and $\lim_{t\to\infty} \dot{\phi}(t)=0$.

(proof) As a candidate of Liapunov function, introduce

$$V = \phi^T(t)\Gamma^{-1}\phi(t). \tag{4.22}$$

Since $\Gamma(0)$ is positive definite, so is $\Gamma^{-1}(0)$. From this and $\Gamma^{-1}(t)=\omega(t)\omega^T(t)$, $\Gamma^{-1}(t)$ is positive definite for all $t\geq 0$. Therefore, V cannot be negative. On the other hand,

$$\begin{aligned} \dot{V} &= 2\phi^T(t)\Gamma^{-1}(t)\dot{\phi}(t) + \phi^T(t)[\Gamma^{-1}]^{\cdot}\phi(t) \\ &= -2\phi^T(t)\omega(t)e(t) + \phi^T(t)\omega(t)\omega^T(t)\phi(t) \\ &= -e^2(t) \leq 0. \end{aligned} \tag{4.23}$$

Therefore, V decreases monotonically. Furthermore, V tends to infinity if and only if $||\phi||$ tends to infinity. Therefore, $\phi(t)$ is uniformly bounded. Also, since V is bounded below, V converges to some nonnegative constant, which implies that $\dot{V}$ converges to zero. Therefore, e(t) converges to zero, which implies that $\phi^T(t)\omega(t)$ converges to zero from eq.(4.10). The argument up to this point holds regardless of the boundedness of $\omega(t)$. With the assumption that $\omega(t)$ is uniformly bounded and the fact that $\Gamma^{-1}(t)$ is bounded, $\dot{\phi}(t)$ converges to zero from eq.(4.21). Q.E.D.

Another useful adaptive law is now presented, while it does not manifest a so fast convergence. The adaptive law is

$$\dot{\tilde{\theta}} = -\Gamma \frac{\omega(t)e(t)}{c+\omega^T(t)\omega(t)}, \quad \Gamma=\Gamma^T>0, \; c>0 \tag{4.24}$$

We donot assume the uniform boundedness of $\omega(t)$ here.

Theorem 4.2 (adaptive law 2) Adaptive law (4.24) yields $\lim_{t\to\infty} e(t)=0$, $\lim_{t\to\infty} \phi^T(t)\omega(t)=0$, and $\lim_{t\to\infty} \dot{\phi}(t)=0$.

(proof) As a candidate of Liapunov function, introduce

$$V = \phi^T(t)\Gamma^{-1}\phi(t).$$

(4.25)

Since Γ is positive definite, V cannot be negative. On the other hand,

$$\begin{aligned}\dot{V} &= 2\phi(t)\Gamma^{-1}\dot{\phi}(t)\\ &= -\frac{2}{c+\omega^T(t)\omega(t)} e^2(t) \leq 0.\end{aligned} \tag{4.26}$$

Therefore, V decreases monotonically. Furthermore, V tends to infinity if and only if $||\phi||$ tends to infinity. Therefore, $\phi(t)$ is uniformly bounded. Also, since V is bounded below, V converges to some nonnegative constant, which implies that $\dot{V}$ converges to zero. Now, if $\omega(t)$ is uniformly bounded, e(t) will converge to zero from eq. (4.26), which implies that $\phi^T(t)\omega(t)$ converges to zero from eq.(4.10) and also that $\dot{\phi}(t)$ converges to zero. On the other hand, assume that $\omega(t)$ is not uniformly bounded. Then, $|e(t)|$ will grow with lower rate than $||\omega(t)||$ from the fact that $\dot{V}$ converges to zero, if $|e(t)|$ would grow to infinity. Suppose that $\phi(t)$ neither tends to zero nor tends to be orthogonal to $\omega(t)$. Then, $|e(t)|$ will grow with the same rate as $||\omega(t)||$, which is contradiction. Hence, either $\phi(t)$ converges to zero or tends to be orthogonal to $\omega(t)$. The former case implies the completion of parameter identification, and the latter case implies that e(t) converges to zero;i.e., either case yields $\lim_{t\to\infty} e(t)=0$ and $\lim_{t\to\infty} \phi^T(t)\omega(t)=0$. From eq.(4.24), it is clear that $\lim_{t\to\infty} e(t)=0$ implies $\lim_{t\to\infty} \dot{\phi}(t)=0$ redardless of zero from eq.(4.10). Furthermore, since $\omega(t)/[c+\omega^T(t)\omega(t)]$ in eq.(4.11) is finite, $\dot{\phi}(t)$ must converge to zero. Q.E.D.

The fact that $\dot{\phi}(t)\to 0$; i.e., $[\theta(t)] \to 0$ does not imply $\tilde{\theta}(t)\to$const., but implies that the motion of vector $\phi(t)$ decreases unlimitedly. The fact that $\phi^T(t)\omega(t)\to 0$ and $\phi^T(t)\Gamma^{-1}\phi(t)\to$ const. means that the vector $\phi(t)$ lies on the ellipsoid $x^T\Gamma^{-1}x$ = const. in 2n dimensional space, being orthogonal to $\omega(t)$. Since the velocity of $\phi(t)$ is infinitesimally small, $\phi(t)$ must be at the origin if $\omega(t)$ spans 2n di-mensional space within a finite time interval. Then, the original objective, i.e., $\phi(t)\to 0$ is achieved. We have seen that the adaptive identi-fication is not always achieved by adaptive law only, but some property of $\omega(t)$ is needed for identification such that $\omega(t)$ spans 2n dimensional space in a finite time interval. Clearly, the property of $\omega(t)$ depends on the property of u(t). The u(t) which bestows $\omega(t)$ such a property is called "sufficiently rich". The necessary and sufficient condition is that there exists a finite time T such that u(t) contains n independent time function on $[t_1, t_1+T]$ for all $t_1 \in [0, \infty)$. Speaking plainly, the condition is that u(t) contains n distinct frequency components.

4.2 Adaptive control. Let the plant and reference model transfer functions be given by

$$t(s) = y(s)/u(s) = gr(s)p^{-1}(s) \qquad (4.27)$$

$$t_d(s) = y_m(s)/v(s) = g_d r_d(s) p_d^{-1}(s) \qquad (4.28)$$

respectively, where $r(s)$, $p(s)$, $r_d(s)$, and $p_d(s)$ are all monic with $\partial[r(s)]=m$, $\partial[p(s)]=n$, $\partial[r_d(s)]=m_d$, and $\partial[p_d(s)] =n_d$. The same assumptions as exact model matching are reqiured; i.e.,

A.1 $r(s)$ is stable.

A.2 $n_d-m_d \geq n-m$.

By introducing arbitrary monic stable polynomials $r^*(s)$ and $p^*(s)$, we provide input dynamics

$$t_{IN} = \frac{r_d(s)p^*(s)}{r^*(s)p_d(s)} . \qquad (4.29)$$

By exact model matching theory explained in Chap. 3, exact model matching is achieved when polynomials $k_b(s)$ and $h(s)$ are determined from the equation

$$k_b(s)p(s)+h(s)r(s)=T(s)r^*(s)p(s)-T(s)r(s)p^*(s) \qquad (4.30)$$

and then the control law is set as

$$u(s) = \frac{1}{g}\{\frac{gk_b(s)}{T(s)r^*(s)}\, u(s)+\frac{h(s)}{T(s)r^*(s)}\, y(s)\}+\frac{g_d}{g}\, \bar{v}(s). \qquad (4.31)$$

Denote $gk_b(s)$ and $h(s)$ as

$$gk_b(s) = k(s) = k_{n-2}s^{n-2}+\cdots+k_1 s+k_0 \qquad (4.32a)$$

and

$$h(s) = h_{n-1}s^{n-1}+h_{n-2}s^{n-2}+\cdots+h_1 s+h_0. \qquad (4.32b)$$

Further, let us define θ and $\omega(t)$ as follows.

$$\theta = [-k_{n-2},\cdots,-k_0,-h_{n-1},\cdots,-h_0]^T, \qquad (4.33)$$

and

$$\omega(t) = [\frac{p^{n-2}}{T(p)r^*(p)}u(t),\cdots,\frac{1}{T(p)r^*(p)}u(t),$$

$$\frac{p^{n-1}}{\tau(p)r^*(p)}y(t),\cdots,\frac{1}{\tau(p)r^*(p)}y(t)]^T. \qquad (4.34)$$

Then the control law for achieving exact model matching can be written as

$$u(t) = -(1/g)\theta^T\omega(t) + (1/g)g_d\bar{v}(t), \qquad (4.35)$$

where θ is a 2n-1 dimensional controller parameter depending on $\{g,r(s),p(s)\}$.

In the case of adaptive control, since $\{g,r(s),p(s)\}$ is unknown, neither is θ. Then, using $\tilde{\theta}(t)$, which is adjustable and represents the estimate of θ, we set the control law for achieving adaptive control as

$$u(t) = -(1/\tilde{g}(t))[\tilde{\theta}(t)]^T\omega(t) + (1/\tilde{g}(t))g_d\bar{v}(t). \qquad (4.36)$$

Although the transfer function (4.27) must represent the plant dynamics, another representation for the plant dynamics using $\{g,k_b(s),h(s)\}$ or $\{g,k(s),h(s)\}$ will be looked for. Dividing both sides of eq.(4.30) by $p(p)\tau(p)\times p^*(p)$ and then multiplying $u(s)$,

$$\frac{k_b(s)}{\tau(s)p^*(s)}u(s) + \frac{h(s)}{\tau(s)p^*(s)}\frac{1}{g}y(s) = \frac{r^*(s)}{p^*(s)}u(s) - \frac{1}{g}y(s)$$

or

$$y(t) = g\frac{r^*(s)}{p^*(p)}u(t)-\frac{k(p)}{\tau(p)r^*(p)}\frac{r^*(p)}{p^*(p)}u(t)$$

$$-\frac{h(p)}{\tau(p)r^*(p)}\frac{r^*(p)}{p^*(p)}y(t),$$

which is further rewritten into

$$y(t) = g\,\frac{r^*(p)}{p^*(p)}u(t) + \theta^T\,\frac{r^*(p)}{p^*(p)}\omega(t), \qquad (4.37)$$

by using the notations of θ and $\omega(t)$. It is to be noticed that eq.(4.37) is a representation of the plant dynamics likewise the transfer function, but is represented by $\{g, \theta\}$ and hence is characterized by not having unknown polynomials in the denominator.

The adaptive control error $e(t)$ is $y(t)-y_m(t)$. Thus

$$e(t) = g\,\frac{r^*(p)}{p^*(p)}u(t) + \theta^T\,\frac{r^*(p)}{p^*(p)}\,\omega(t) - y_m(t), \qquad (4.38)$$

which is an error dynamics for adaptive control. Although the objective of adaptive control is to make $e(t)$ zero, let us regard the error dynamics just as if an unknown plant in the adaptive identification problem. Then the identifier for adaptive control error dynamics becomes

$$\tilde{e}(t) = \tilde{g}(t)\frac{r^*(p)}{p^*(p)}u(t) + \tilde{\theta}^T(t)\frac{r^*(p)}{p^*(p)}\omega(t) - y_m(t). \qquad (4.39)$$

The adjustable parameters, $\tilde{g}(t)$ and $\tilde{\theta}(t)$ employed in the control law (4.36) are meant by $\tilde{g}(t)$ and $\tilde{\theta}(t)$ defined here. Denoting identification error $\tilde{e}(t)-e(t)$ by $\varepsilon(t)$, we have

$$\begin{aligned}\varepsilon(t) &= (\tilde{g}(t)-g)\frac{r^*(p)}{p^*(p)}u(t) + [\tilde{\theta}^T(t)-\theta^T]\frac{r^*(p)}{p^*(p)}\omega(t) \\ &= [\phi_g(t), \phi^T(t)]\begin{bmatrix}\frac{r^*(p)}{p^*(p)}\,u(t) \\ \frac{r^*(p)}{p^*(p)}\,\omega(t)\end{bmatrix}, \end{aligned} \qquad (4.40)$$

where $\phi_g(t) = \tilde{g}(t) - g$ and $\phi(t) = \tilde{\theta}(t) - \theta$. Notice that eq.(4.40) is an identification error dynamics. Following Theorem 4.2, let us use the following adaptive law;

$$\begin{bmatrix} \tilde{g}(t) \\ \tilde{\theta}(t) \end{bmatrix}^{\cdot} = -\Gamma \frac{\begin{bmatrix} \frac{r^*(p)}{p^*(p)}u(t) \\ \frac{r^*(p)}{p^*(p)}\omega(t) \end{bmatrix} \varepsilon(t)}{c + \begin{bmatrix} \frac{r^*(p)}{p^*(p)}u(t) \\ \frac{r^*(p)}{p^*(p)}\omega(t) \end{bmatrix}^T \begin{bmatrix} \frac{r^*(p)}{p^*(p)}u(t) \\ \frac{r^*(p)}{p^*(p)}\omega(t) \end{bmatrix}}, \quad (4.41)$$

where $\Gamma=\Gamma^T>0$ and $c>0$. It will be shown that

$$\left.\begin{aligned} &\lim_{t\to\infty} \varepsilon(t) = 0 \\ &\lim_{t\to\infty} [\phi_g(t) \quad \phi^T(t)] \begin{bmatrix} r^*(p)/p^*(p)\cdot u(t) \\ r^*(p)/p^*(p)\cdot \omega(t) \end{bmatrix} = 0 \\ &\lim_{t\to\infty} \dot{\phi}_g(t) = 0, \ \lim_{t\to\infty} \dot{\phi}(t) = 0 \end{aligned}\right\} \quad (4.42)$$

holds, from which the total stability of the adaptive control system is concluded, as will be proven in the following.

The structure of this adaptive control system is essentially the same as that developed by Narendra [4] or that developed by the author [14], but does not need introduction of the concept of positive realness nor the concept of augmented error signal.

For brevity, let us introduce the notation $\Phi(t)$ and $\Omega(t)$ by

$$\Phi(t) = [\phi_g(t) \quad \phi^T]^T \quad (4.30)$$

$$\Omega(t) = [\frac{r^*(p)}{p^*(p)}u(t) \quad \frac{r^*(p)}{p^*(p)}\omega^T(t)]^T. \quad (4.31)$$

Then eqns.(4.40) and (4.41) are rewritten as follows.

$$\varepsilon(t) = \Phi^T(t)\Omega(t) \tag{4.45}$$

$$\dot{\Phi}(t) = -\Gamma \frac{\Omega(t)\varepsilon(t)}{c + \Omega^T(t)\Omega(t)}. \tag{4.46}$$

Introduce a candidate of Liapunov function by

$$V(\Phi(t)) = \Phi^T(t)\Gamma^{-1}\Phi(t), \tag{4.47}$$

where V is nonnegative and tends to infinity only when $||\Phi(t)|| \to \infty$. The time rate of change of V is evaluated as

$$\dot{V}(\Phi(t)) = -2\ \varepsilon^2(t)/[c+\Omega^T(t)\Omega(t)] \leq 0 \tag{4.48}$$

Since V decreases monotonically, $\Phi(t)$ is always bounded. Also, since V decreases monotonically and bounded below, V must converge to some constant and hence $\dot{V}$ converges to zero. That is,

$$\lim_{t\to\infty} \frac{\varepsilon^2(t)}{c + \Omega^T(t)\Omega(t)} = 0. \tag{4.49}$$

On the assumption that $||\Omega(t)||$ is bounded, $\lim_{t\to\infty} \varepsilon(t) = 0$ is obtained and hence $\lim_{t\to\infty} \dot{\Phi}(t) = 0$ follows from eq.(4.45) since $\Omega(t)/[c+\Omega^T(t)\Omega(t)]$ is bounded. However, the boundedness of $||\Omega(t)||$ is not assured yet. Then the diverging velocity of $|\varepsilon(t)|$, if it would diverge, is of lower order than that of $||\Omega(t)||$ because of eq.(4.49), and hence $\lim_{t\to\infty} \dot{\Phi}(t) = 0$ follows again from eq.(4.46). Let us now examine eq.(4.45). If $\Phi(t)$ is nonzero and does not orthogonalize with $\Omega(t)$, then $|\varepsilon(t)|$ will diverge with the same velocity as $||\Omega(t)||$, which is the contradiction. Thus, it follows that either $\Phi(t) \to 0$ or that $\Phi(t)$ orthogonalizes with $\Omega(t)$ with being not necessarily zero and hence $\varepsilon(t) \to 0$ holds. The former case means the convergence of the adjustable parameters to those for exact model matching and hence the completion of adaptive control scheme is yielded. Then the only question is the latter case.

Summarizing above, it can be said that although nothing can be said about the bondedness of $||\Omega(t)||$, $\varepsilon(t) \to 0$ and $\dot{\Phi}(t) \to 0$ hold together with $\Phi(t)^T\Omega(t) \to 0$ of course. Now a lemma is provided in order to show that $\lim_{t\to\infty} \tilde{e}(t)=0$, the convergence of u(t) to the plant input function in exact model matching, and the boundedness of $\Omega(t)$.

Lemma 4.1. Let f(t) be any differentiable time function. If $[\tilde{g}(t)]^{\cdot}=0$, then the followings hold.

1) $$\lim_{t\to\infty} r^*(p)[\tilde{g}(t)f(t)] = \lim_{t\to\infty} \tilde{g}(t)r^*(p)f(t) \tag{4.50}$$

2) $$\lim_{t\to\infty} \frac{1}{p^*(p)}[\tilde{g}(t)f(t)] = \lim_{t\to\infty} \tilde{g}(t)\frac{1}{p^*(p)}f(t) \tag{4.51}$$

3) $$\lim_{t\to\infty} \frac{r^*(p)}{p^*(p)}[\tilde{g}(t)f(t)] = \lim_{t\to\infty} \tilde{g}(t)\frac{r^*(p)}{p^*(p)}f(t) \tag{4.52}$$

(proof)

1) $r^*(p)[\tilde{g}(t)f(t)]$

$= (r_0^* + r_1^*p + \cdots + r_m^*p^m)[\tilde{g}(t)f(t)]$

$= \tilde{g}(t)(r_0^*f(t)+r_1^*pf(t)+\cdots+r_m^*p^mf(t))$+(derivative terms of $\tilde{g}(t)$)

$= \tilde{g}(t)r^*(p)f(t)$ + (derivative terms of $\tilde{g}(t)$)

However, since $\lim_{t\to\infty} [\tilde{g}(t)] = 0$, eq.(4.37) results.

2) $\lim_{t\to\infty} p^*(p)[\tilde{g}(t)\frac{1}{p^*(p)}f(t)]$

$$= \lim_{t\to\infty} \tilde{g}(t)[p^*(p)\frac{1}{p^*(p)}f(t)] \quad \text{(by 1))}$$

$$= \lim_{t\to\infty} \tilde{g}(t)f(t)$$

Thus eq.(4.51) holds.

3) $$\lim_{t\to\infty} \tilde{g}(t)\frac{r^*(p)}{p^*(p)}f(t)$$

$$= \lim_{t\to\infty} \frac{1}{p^*(p)}\tilde{g}(t)r^*(p)f(t) \quad \text{(by 2))}$$

$$= \lim_{t\to\infty} \frac{1}{p^*(p)}r^*(p)[\tilde{g}(t)f(t)] \quad \text{(by 1))}$$

$$= \lim_{t\to\infty} \frac{r^*(p)}{p^*(p)}[\tilde{g}(t)f(t)]$$ Q.E.D.

This lemma is clearly extended to the case when both $\tilde{g}(t)$ and $f(t)$ are vectors. Therefore, from the fact that $[\tilde{g}(t)]^{\cdot} \to 0$ and $[\tilde{\vartheta}(t)]^{\cdot} \to 0$, we obtain

$$\lim_{t\to\infty} \tilde{g}(t)\frac{r^*(p)}{p^*(p)}u(t) = \lim_{t\to\infty} \frac{r^*(p)}{p^*(p)}\tilde{g}(t)u(t) \tag{4.53}$$

and

$$\lim_{t\to\infty} \tilde{\vartheta}^T(t)\frac{r^*(p)}{p^*(p)}\omega(t) = \lim_{t\to\infty} \frac{r^*(t)}{p^*(p)}\tilde{\vartheta}^T(t)\omega(t). \tag{4.54}$$

Then, from eq.(4.39) we obtain

$$\lim_{t\to\infty} \tilde{e}(t) = \lim_{t\to\infty} \frac{r^*(p)}{p^*(p)}\{\tilde{g}(t)u(t)+\tilde{\vartheta}^T(t)\omega(t)\}-y_m(t),$$

which implies, in view of eq.(4.36),

$$\lim_{t\to\infty} \tilde{e}(t) = 0. \tag{4.55}$$

Since $\lim_{t\to\infty} \varepsilon(t)=0$ has already been obtained, we finally obtain

$$\lim_{t\to\infty} e(t) = 0.$$

Furthermore, $\bar{\Phi}^T(t)\Omega(t) \to 0$ can be rewritten as

$$\lim_{t\to\infty} \{(\tilde{g}(t)-g)\frac{r^*(p)}{p^*(p)}u(t)+(\tilde{\theta}^T(t)-\theta^T)\frac{r^*(p)}{p^*(p)}\omega(t)\}=0,$$

which is, by Lemma 4.1, further rewritten as

$$\lim_{t\to\infty} \frac{r^*(p)}{p^*(p)} \{(\tilde{g}(t)-g)u(t) + (\tilde{\theta}^T(t)-\theta^T)\omega(t)\} = 0.$$

Since $r^*(p)/p^*(p)$ is a stable filter, the above equation implies

$$\lim_{t\to\infty} \{(\tilde{g}(t)-g)u(t) + (\tilde{\theta}^T(t)-\theta^T)\omega(t)\} = 0,$$

from which, by employing eq.(4.36), we finally obtain

$$\lim_{t\to\infty} u(t) = \lim_{t\to\infty} \{ -\frac{1}{g}\,\theta^T\omega(t) + \frac{g_d}{g}\,\bar{v}(t)\}. \qquad (4.56)$$

The last relation implies that the input to the plant converges to the input function that would be applied to the plant in exact model matching while executing adaptive control, where the input is assured to be bounded from Theorem 3.2. Thus it has been concluded that $\Omega(t)$ is bounded, i.e., all signals in the adaptive control system are bounded, and $\lim_{t\to\infty} e(t)=0$ is achieved.

REFERENCES

[1] I.D.Landau; A survey of model reference adaptive techniques ——— theory and applications, Automatica 10, 353/379,1974.

[2] R.V.Monopoli; Model reference adaptive control with an augmented error signal, IEEE Trans. Vol.AC-19,474/484, 1974.

[3] A.Feuer and A.S.Morse; Adaptive control of single-input, single-output linear systems, IEEE Trans. Vol. AC-23,557/569,1978.

[4] K.S.Narendra and L.S.Valavani; Stable adaptive controller design —— Direct control, ibid,570/583,1978.

[5] G.C.Goodwin,R.J.Ramadge and P.E.Caines; Discrete time multivariable adaptive control, IEEE Trans. Vol.AC-25, 449/456,1980

[6] Y.D.Landau; Adaptive control, Dekker,1979.

[7] B.Egardt; Stability of adaptive control, Springer-Verlag,1979.

[8] K.S.Narendra and R.V.Monopoli(ed.); Applications of adaptive control, Academic Press,1980.

[9] I.D.Landau and M.Tomizuka; Theory and practice of adaptive control systems (in Japanese), Ohm,1981.

[10] K.Ichikawa,K.Kanai,T.Suzuki, and K.Tamura; Adaptive control (in Japanese), Shokodo,1984.

[11] S.Kobayashi; Model reference adaptive control system, Keisoku to Seigyo 17,602/612 (in Japanese),1978.

[12] T.Suzuki; Adaptive observer, ibid, 19, 195/205 (in Japanese),1980.

[13] K.Tamura and K.Ichikawa; Design of model reference adaptive control system, ibid, 20, 1078/1086 (in Japnese),1981.

[14] K.Ichikawa; Continuous time adaptive identification and control algorithms via newly developed adaptive laws, Int. J. Control,36,819/831,1982.

Chapter 5 Disturbance in exact model matching and adaptive control

In the preceding two chapters, disturbance was not taken into consideration. In the first, the method to remove the effect of known disturbance in the exact model matching is developed. Then, it is extended to remove the effect of unknown disturbance in the adaptive control.

5.1 The effect of disturbance. Two kinds of disturbance are thought to exist in control systems. The first is the one superimposed on the control input signal u(t); the second is the one superimposed on the output signal y(t), called observation noise. We consider the first type of disturbance only here. The plant with disturbance is shown in Fig. 5.1. The plant dynamics is described by

$$y(s) = \frac{gr(s)}{p(s)} u(s) + \frac{g_w r_w(s)}{p(s)} w(s), \tag{5.1}$$

where w(s) is disturbance, p(s) is an n degree monic polynomial, and r(s) and $r_w(s)$ are m and m_w degree monic polynomials respectively, with $0 \leq (m, m_w) \leq n-1$. It is also assumed that r(s) and p(s) are relatively prime, with r(s) being stable. The control law to achieve exact model matching without disturbance was found on Chapter 3 such as

$$u(s) = \frac{1}{g}\{\frac{k(s)}{\gamma(s)r^*(s)} u(s) + \frac{h(s)}{\gamma(s)r^*(s)} y(s)\} + \frac{g_d}{g} \bar{v}(s). \tag{5.2}$$

When disturbance is not applied to the plant, the plant output y(t) converges to the reference model output $y_m(t)$. The plant output y(t), however, becomes

$$y(s) = y_m(s) + \frac{g_w\{h(s)+\gamma(s)p^*(s)\}r_w(s)}{\gamma(s)p^*(s)p(s)} w(s), \tag{5.3}$$

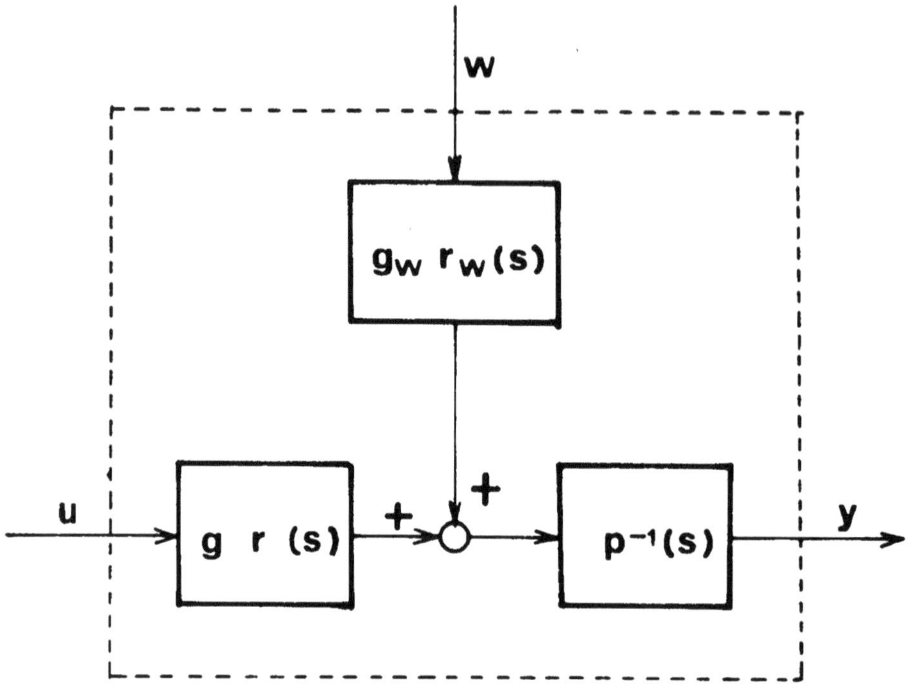

Fig. 5.1 Block diagram of the plant with disturbance

where the second term represents error(s) caused by disturbance. From the polynomial equation derived from eq.(3.11) , we obtain

$$\frac{1}{g}\,k(s) - \gamma(s)r^*(s) = -\,\frac{r(s)\{h(s)+\gamma(s)p^*(s)\}}{p(s)}. \qquad (5.4)$$

Since $r(s)$ and $p(s)$ are relatively prime, $h(s)+\gamma(s)p^*(s)$ must be divided by $p(s)$. Clearly, $h(s)$ is of strictly lower degree than $p(s)$ and $\gamma(s)p^*(s)$ is of no less degree than $p(s)$. Divide $\gamma(s)p^*s)$ by $p(s)$ and denote the quotient and residual by $\alpha(s)$ and $r_e(s)$ respectively. It is clear that $h(s)$ must be equal to $-r_e(s)$ and $\{h(s)+\gamma(s)p^*(s)\}/p(s)$ becomes to $\alpha(s)$. Therefore, $e(s)$, the error due to disturbance, can be described as

$$e(s) = \frac{g_w\alpha(s)r_w(s)}{\gamma(s)p^*(s)}\,w(s). \qquad (5.5)$$

Since $\gamma(s)p^*(s)$ is stable, $e(t)$ is bounded for bounded disturbance $w(t)$. Also, when $w(t)$ contonues to be zero, $e(t)\to 0$ as $t\to\infty$. In the following, the controller is modified so that $e(t)\to 0$ even if $w(t)$ is applied to the plant persistently.

5.2 Suppression of disturbance effect in exact model matching. The disturbance is either unknown nor unavailable in nature, but we assume here the knowledge and availability of disturbance. The unknown disturbance will be dealt with in context to adaptive control.

Theorem 5.1 The control law

$$u(s)=\frac{1}{g}\{\frac{k(s)}{\gamma(s)r^*(s)}u(s) + \frac{h(s)}{\gamma(s)r^*(s)}y(s)\} + \frac{g_d}{g}\,\bar{v}(s)$$

$$-\frac{g_w}{\text{[illegible]}}\cdot\frac{\alpha(s)r_w(s)}{\text{[illegible]}}\,w(s) \qquad (5.6)$$

not only acieves exact model matching but also suppresses the effect of disturbance.

(proof) From the polynomial equation, we obtain

$$\frac{1}{g}\frac{k(s)}{\tau(s)r^*(s)}u(s)+\frac{h(s)}{\tau(s)r^*(s)}\frac{r(s)}{p(s)}u(s)=u(s)$$
$$-\frac{r(s)}{p(s)}u(s)\frac{p^*(s)}{r^*(s)}. \tag{5.7}$$

However, since $r(s)p^{-1}(s)u(s)=(1/g)y(s)-(g_w/g)r_w(s)p^{-1}(s)$ $\times w(s)$, it becomes

$$\frac{1}{g}\frac{k(s)}{\tau(s)r^*(s)}u(s)+\frac{1}{g}\frac{h(s)}{\tau(s)r^*(s)}y(s)-\frac{g_w}{g}\frac{h(s)}{\tau(s)r^*(s)}\frac{r_w(s)}{p(s)}$$
$$\times w(s)=u(s)-\frac{1}{g}y(s)\frac{p^*(s)}{r^*(s)}+\frac{g_w}{g}\frac{r_w(s)}{p(s)}w(s)\frac{p^*(s)}{r^*(s)}. \tag{5.8}$$

From eqs.(5.6) and (5.8), we obtain

$$\frac{1}{g}\frac{p^*(s)}{r^*(s)}y(s)-\frac{g_w}{g}\frac{r_w(s)}{p(s)}\left\{\frac{h(s)}{\tau(s)r^*(s)}+\frac{p^*(s)}{r^*(s)}\right\}w(s)$$
$$=\frac{g_d}{g}\bar{v}(s)-\frac{g_w}{g}\frac{\alpha(s)r_w(s)}{\tau(s)r^*(s)}w(s). \tag{5.9}$$

However, since $\{h(s)+\tau(s)p^*(s)\}/p(s)=\alpha(s)$, we obtain

$$y(s)=\frac{g_d r^*(s)}{p^*(s)}\bar{v}(s). \tag{5.10}$$

Furthermore, since $r(s)$ is assumed to be stable, $u(t)$ is bounded for bounded $\bar{v}(t)$. Q.E.D.

5.3 Representation of disturbance and its generation.

We consider two kinds of disturbance as persistently applied disturbance. The first is polynomial form disturbance, and th second is multiple frequency sinusoidal disturbance.

5.3.1 Polynomial form disturbance.

Polynomial form disturbance of degree d is represented as

$$w(t)=w_d(t^d/d!)+\cdots\cdots+w_1t+w_0. \qquad (5.11)$$

We cosider that w(t) is generated as an impulse response of a certain linear system. Since the Laplace transform of an impulse response is equal to the system transfer function, the system transfer function will be

$$t_w(s) = \frac{w_0 s^d + w_1 s^{d-1} + \cdots\cdots + w_d}{s^{d+1}}. \qquad (5.12)$$

5.3.2 Sinusoidal disturbance. Superimposed wave form of d distinct sine waves is represented as

$$w(t) = \sum_{i=1}^{d} (a_i \cos \omega_i t + b_i \sin \omega_i t). \qquad (5.13)$$

The system transfer function which generates w(t) as its own impulse response is

$$t_w(s) = \sum_{i=1}^{d} [(a_i/\omega_i)s+b_i]/(s^2+\omega_i^2). \qquad (5.14)$$

5.4 Control law for the specified disturbance.

5.4.1 Polynomial form disturbance. The disturbance is represented by eq.(5.12). Denote the polynomial $g_w\alpha(s) \times r_w(s)(w_0 s^d+w_1 s^{d-1}+\cdots+w_d)$ by K(s), the degree of which is $\mu=n-m-1+m_w+d$. Then, the control law (5.6) is rewritten as

$$u(s) = \frac{1}{g}\{\frac{k(s)}{\tau(s)r^*(s)} u(s) + \frac{h(s)}{\tau(s)r^*(s)} y(s)\} + \frac{g_d}{g} \bar{v}(s)$$
$$- \frac{1}{g} \frac{K(s)}{\tau(s)r^*(s)} \frac{1}{s^{d+1}}. \qquad (5.15)$$

Now, put

$$\left.\begin{aligned} k(s) &= \qquad\qquad k_{n-2}s^{n-2}+\cdots\cdots+k_0 \\ h(s) &= h_{n-1}s^{n-1}+h_{n-2}s^{n-2}+\cdots\cdots+h_0 \\ K(s) &= K_\mu s^\mu+\cdots\cdots\cdots\cdots\cdots\cdots+K_0 \end{aligned}\right\}. \qquad (5.16)$$

Define

$$\left.\begin{aligned}\theta &= [-k_{n-2},\cdots,-k_0,-h_{n-1},\cdots,-h_0]^T \\ \phi &= [K_\mu,\cdots\cdots.K_0]^T\end{aligned}\right\}. \tag{5.17}$$

Also, define

$$\omega(t) = [\frac{p^{n-2}}{\gamma(p)r^*(p)}u(t),\cdots,\frac{1}{\gamma(p)r^*(p)}u(t), \\ \frac{p^{n-1}}{\gamma(p)r^*(p)}y(t),\cdots,\frac{1}{\gamma(p)r^*(p)}y(t)]^T. \tag{5.18}$$

Let us denote the impulse response of the system $s^{-(d+1)}$ by $f(t)$, and define

$$\zeta(t) = [\frac{p^\mu}{\gamma(p)r^*(p)}f(t),\cdots,\frac{1}{\gamma(p)r^*(p)}f(t)]^T. \tag{5.19}$$

Then, the control law (5.15) is expressed as

$$u(t) = -\frac{1}{g}\,\theta^T\omega(t) + \frac{g_d}{g}\,\bar{v}(t) - \frac{1}{g}\,\phi^T\zeta(t). \tag{5.20}$$

5.4.2 Sinusoidal disturbance. $w(s)$ is represented by eq.(5.14). Denote the polynomial $g_w\alpha(s)w(s)/\{(a_i/\omega_i)s+b_i\}$ by $H_i(s)$, the degree of which is $\nu=n-m-1+m_w+1$. Put

$$H_i(s) = H_{i,\nu}s^\nu + \cdots + H_{i,0},\ i=1,2,\cdots,d \tag{5,21}$$

and define a $(\nu+1)\cdot d$ parameter vector ψ by

$$\psi = [H_{1,\nu},\cdots,H_{1,0},\cdots\cdots,H_{d,\nu},\cdots,H_{d,0}]^T. \tag{5.22}$$

Denote the impulse response of the system $1/(s^2+\omega_i{}^2)$ by $f_i(t)$, and define a$(\nu+1)\cdot d$ signal vector $\xi(t)$ by

$$\xi(t) = [\frac{p^\nu}{\gamma(p)r^*(p)}f_1(t),\cdots,\frac{1}{\gamma(p)r^*(p)}f_1(t),\cdots,$$

$$\frac{p^{\nu}}{\tau(p)r^{*}(p)}f_d(t),\cdots,\frac{1}{\tau(p)r^{*}(p)}f_d(t)]^T. \qquad (5.23)$$

Then, the control is expressed as

$$u(t) = -\frac{1}{g}\,\theta^T\omega(t) + \frac{g_d}{g}\,\bar{v}(t) - \frac{1}{g}\,\phi^T\xi(t). \qquad (5.24)$$

5.5 Suppression of disturbance effct in adaptive control.

5.5.1 Polynomial form disturbance. In adaptive control, g, θ and ϕ are unknown. Therefore, the control law is set as

$$u(t) = -\frac{1}{\tilde{g}(t)}\,\tilde{\theta}^T(t)\omega(t)+\frac{g_d}{\tilde{g}(t)}\,\bar{v}(t)-\frac{1}{\tilde{g}(t)}\tilde{\phi}^T(t)\zeta(t). \qquad (5.25)$$

From the polynomial equation and eq.(5.1), we obtain

$$y(s) = -\frac{k(s)}{\tau(s)p^{*}(s)}u(s)-\frac{h(s)}{\tau(s)p^{*}(s)}y(s)+g\frac{r^{*}(s)}{p^{*}(s)}u(s)$$
$$+\frac{h(s)+\tau(s)p^{*}(s)}{\tau(s)p^{*}(s)}\cdot\frac{g_w r_w(s)}{p(s)}w(s). \qquad (5.26)$$

However, since $\{h(s)+\tau(s)p^{*}(s)\}/p(s)$ is $\alpha(s)$, it reduces to

$$y(s) = -\frac{k(s)}{\tau(s)p^{*}(s)}u(s)-\frac{h(s)}{\tau(s)p^{*}(s)}y(s)+g\frac{r^{*}(s)}{p^{*}(s)}u(s)$$
$$+\frac{g_w\alpha(s)r_w(s)}{\tau(s)p^{*}(s)}w(s), \qquad (5.27)$$

which is further written as

$$y(s) = -\frac{k(s)}{\tau(s)r^{*}(s)}\,\frac{r^{*}(s)}{p^{*}(s)}u(s)-\frac{h(s)}{\tau(s)r^{*}(s)}\,\frac{r^{*}(s)}{p^{*}(s)}y(s)$$
$$+g\frac{r^{*}(s)}{p^{*}(s)}u(s)+\frac{g_w\alpha(s)r_w(s)}{\tau)s)r^{*}(s)}\,\frac{r^{*}(s)}{p^{*}(s)}w(s). \qquad (5.28)$$

Using the notations of θ, ϕ and $\zeta(t)$, eq.(5.28) is expressed as

$$y(t) = \theta^T \frac{r^*(p)}{p^*(p)}\omega(t)+g\frac{r^*(p)}{p^*(p)}u(t)+\phi^T \frac{r^*(p)}{p^*(p)}\zeta(t), \qquad (5.29)$$

which is the extension of eq.(4.37). The adaptive control error $e(t)=y(t)-y_m(t)$ is then

$$e(t)=\theta^T \frac{r^*(p)}{p^*(p)}\omega(t)+g\frac{r^*(p)}{p^*(p)}u(t)+\phi^T \frac{r^*(p)}{p^*(p)}\zeta(t)-y_m(t). \quad (5.30)$$

The identifier is defined by

$$\tilde{e}(t) = \tilde{\theta}^T(t)\frac{r^*(p)}{p^*(p)}\omega(t)+\tilde{g}(t)\frac{r^*(p)}{p^*(p)}u(t)+\tilde{\phi}^T(t)\frac{r^*(p)}{p^*(p)}\zeta(t)$$
$$-y_m(t). \qquad (5.31)$$

The control parameter $\tilde{g}(t)$, $\tilde{\theta}(t)$ and $\tilde{\phi}(t)$ used in eq.(5.25) are the estimates defined by eq.(5.31). The identification error $\varepsilon_p(t)=\tilde{e}(t)-e(t)$ is expressed as

$$\varepsilon_p(t) = [\tilde{\theta}^T(t)-\theta,\tilde{g}(t)-g,\tilde{\phi}(t)-\phi]\begin{pmatrix} r^*(p)/p^*(p)\cdot\omega(t) \\ r^*(p)/p^*(p)\cdot u(t) \\ r^*(p)/p^*(p)\cdot\zeta(t) \end{pmatrix}. \quad (5.32)$$

The problem has now been completely reduced to the conventional adaptive control problem. Let us define $\Omega_p(t)$ by

$$\Omega_p(t)=[\frac{r^*(p)}{p^*(p)}\omega^T(t), \frac{r^*(p)}{p^*(p)}u(t), \frac{r^*(p)}{p^*(p)}\zeta^T(t)]^T. \qquad (5.33)$$

Then, the adaptive law is obtained as

$$\begin{pmatrix} \dot{\tilde{\theta}}(t) \\ \dot{\tilde{g}}(t) \\ \dot{\tilde{\phi}}(t) \end{pmatrix} = -\Gamma_p \frac{\Omega_p(t)\varepsilon_p(t)}{c_p+\Omega_p{}^T(t)\Omega_p(t)}, \Gamma_p=\Gamma_p^T>0, c_p>0. \qquad (5.34)$$

5.5.2 Sinusoidal disturbance. Quite the analogous arguments can be done as polynomial form disturbance. The control law is

$$u(t) = -\frac{1}{\tilde{g}(t)}\,\tilde{\theta}^T(t)\omega(t)+\frac{g_d}{\tilde{g}(t)}\,\bar{v}(t)-\frac{1}{\tilde{g}(t)}\tilde{\psi}^T(t)\xi(t). \quad (5.35)$$

The plant dynamics corresponding to eq.(5.29) is

$$y(t) = \theta^T\,\frac{r^*(p)}{p^*(p)}\omega(t)+g\frac{r^*(p)}{p^*(p)}u(t)+\psi^T\,\frac{r^*(p)}{p^*(p)}\xi(t). \quad (5.36)$$

The identification error is expressed as

$$\varepsilon_s(t) = [\tilde{\theta}(t)-\theta,\tilde{g}(t)-g,\tilde{\psi}(t)-\psi]\begin{bmatrix} r^*(p)/p^*(p)\cdot\omega(t) \\ r^*(p)/p^*(p)\cdot u(t) \\ r^*(p)/p^*(p)\cdot\xi(t)\end{bmatrix}. \quad (5.37)$$

Define $\Omega_s(t)$ by

$$\Omega_s(t)=[\frac{r^*(p)}{p^*(p)}\omega^T(t),\ \frac{r^*(p)}{p^*(p)}u(t),\ \frac{r^*(p)}{p^*(p)}\xi(t)]^T. \quad (5.38)$$

Then, the adaptive law is obtained as

$$\begin{bmatrix}\dot{\tilde{\theta}}(t) \\ \dot{\tilde{g}}(t) \\ \dot{\tilde{\psi}}(t)\end{bmatrix} = -\,\Gamma_s\,\frac{\Omega_s(t)\varepsilon_s(t)}{c_s+\Omega_s{}^T(t)\Omega_s(t)},\Gamma_s=\Gamma_s^T>0,c_s>0. \quad (5.39)$$

Chapter 6 Adaptive pole assignment

As was mentioned before, minimum phase of the plant is inevitable for both exact model matching and adaptive control, because the plant numerator is cancelled by the controller. When the plant is of nonminimum phase and moreover unstable, the only thing one can do for controlling the plant is pole assignment. Clearly, the pole assignment is a subproblem of exact model matching, when the plant is completely known. When, however, the plant is unknown, the situation is converse; i.e., adaptive pole assignment is no longer a subproblem of the conventional adaptive control. There have been many researches on adaptive pole assignment [1]-[5]. Most of them are of indirect method which is simple in principle. Only [5] seems to be of direct method, but it relies on Bezout identity. A new direct method is presented in this chapter. Discrete time systems are considered, the detail treatment of which is illustrated in Chap. 9.

6.1 Problem statement. Let the plant transfer function be given by

$$t(z) \triangleq \frac{y(z)}{u(z)} = \frac{r(z)}{p(z)}, \tag{6.1}$$

where $p(z)$ is n degree monic polynomial and $r(s)$ is of less degree than n. It is also assumed that $r(z)$ and $p(z)$ are relatively prime. Let the desired n degree monic polynomial as the closed loop characyeristic polynomial be $p_d(z)$. The object of adaptive pole assignment is to construct an adaptive system so that the closed loop system transfer function converges to the desired transfer function

$$t_d \triangleq \frac{y_m(z)}{v(z)} = \frac{r(z)}{p_d(z)}. \tag{6.2}$$

Since $r(z)$ is unknown, neither reference model nor its output $y_m(k)$ are unknown.

6.2 Nonadaptive pole assignment. In the first, we consider the nonadaptive problem. Let $q(z)$ be any n-1 degree monic stable polynomial. Then the polynomial equation

$$k(z)p(z) + h(z)r(z) = q(z)[p(z)-p_d(z)] \tag{6.3}$$

has unique solution for the unknown polynomials $k(z)$ and $h(z)$ with degree of at most n-2 and n-1 respectively [6]. The control law

$$u(z) = \frac{k(z)}{q(z)}u(z) + \frac{h(z)}{q(z)}y(z) + v(z) \tag{6.4}$$

achieves pole assignment. Solving eq.(6.3) for $k(z)$ and $h(z)$ is essentially the same as solving a 2n-1 dimensional simultaneous equation.

We demonstrate that solving simultaneous equation can be avoided by adaptively updating the control parameters. Let us put $k(z)$ and $h(z)$ as follows.

$$k(z) = k_{n-2}z^{n-2} + \cdots + k_0 \tag{6.5}$$

$$h(z) = h_{n-1}z^{n-1} + \cdots + h_0 \tag{6.6}$$

Define

$$\theta = [-k_{n-2},\cdots,-k_0,-h_{n-1},\cdots,-h_0]^T, \tag{6.7}$$

and

$$\xi(k) = [\frac{z^{n-2}}{q(z)}\, u(k),\cdots,\frac{1}{q(z)}\, u(k),$$
$$\frac{z^{n-1}}{q(z)}\, y(k),\cdots,\frac{1}{q(z)}\, y(k)]^T. \tag{6.8}$$

Then, the control law (6.4) can be written as

$$u(k) = -\theta^T\xi(k) + v(k). \tag{6.9}$$

Let the estimate of θ defined later be denoted by $\hat{\theta}(t)$. The control law is modified to

$$u(k)= -\tilde{\theta}^T(k)\xi(k) + v(k). \tag{6.10}$$

We easily obtain from eq.(6.3) the relation

$$y(k) = \frac{r(z)}{p_d(z)} u(k)- \frac{k(z)}{q(z)} \frac{r(z)}{p_d(z)} u(k)- \frac{h(z)}{q(z)} \frac{r(z)}{p_d(z)} y(k), \tag{6.11}$$

which can be written by using the notations θ and $\xi(k)$ as

$$y(k) = \frac{r(z)}{p_d(z)} u(k) + \theta^T \frac{r(z)}{p_d(z)} \xi(k). \tag{6.12}$$

The above equation is nothing but the plant dynamics in terms of $r(z)$, $p_d(z)$ and θ. The identifier will then be defined by introducing the estimate $\tilde{\theta}(k)$; i.e.,

$$\tilde{y}(k) = \frac{r(z)}{p_d(z)} u(k) + \tilde{\theta}^T(k-1) \frac{r(z)}{p_d(z)} \xi(k). \tag{6.13}$$

The identification error $\varepsilon(k)=\tilde{y}(k)-y(k)$ can be written as

$$\varepsilon(k) = [\tilde{\theta}(k-1)-\theta]^T \frac{r(z)}{p_d(z)} \xi(k). \tag{6.14}$$

It is known that $\tilde{\theta}(k)$ can be approached to θ by specifying an appropriate adaptive law: for example, we can use

$$\tilde{\theta}(k) = \tilde{\theta}(k-1)- \frac{r(z)p_d^{-1}(z)\xi(k)\varepsilon(k)}{c+[r(z)p_d^{-1}(z)\xi(k)]^T[r(z)p_d^{-1}(z)\xi(k)]}, \quad c>0 \tag{6.15}$$

6.3 Plant identification. Although we are seeking an adaptive system of direct method, the plant identification process is inevitable. We can easily obtain the parametric representation of the plant dynamics as

$$y(k) = \phi^T\zeta(k), \tag{6.16}$$

where

$$\phi = [-p_{n-1},\cdots,-p_0,-r_{n-1},\cdots,-r_0]^T, \tag{6.17a}$$

and

$$\zeta(k) = [y(k-1),\cdots,y(k-n),u(k-1),\cdots,u(k-n)]^T. \tag{6.17b}$$

The identifier is defined by

$$\tilde{y}(k) = \tilde{\phi}^T(k-1)\zeta(k). \tag{6.18}$$

The identification error $\varepsilon_p(k)=\tilde{y}(k)-y(k)$ can be written as

$$\varepsilon_p(k) = [\tilde{\phi}(k-1)-\phi]^T\zeta(k). \qquad (6.19)$$

An appropriate adaptive law is used to update $\tilde{\phi}(k)$; for example,

$$\tilde{\phi}(k) = \tilde{\phi}(k-1) - \frac{\zeta(k)\varepsilon_p(k)}{c_p+\zeta^T(k)\zeta(k)},\ c_p>0. \qquad (6.20)$$

The estimate $\tilde{\phi}(k)$ at time k gives the estimates $\bar{p}(z,k)$ and $\bar{r}(z,k)$. On the assumption of persistent exciting, these estimates converge to p(z) and r(z) respectively.

We consider that at time k the plant dynamics is just $\bar{r}(z,k)\bar{p}(z,k)^{-1}$. Clearly, there exists a control parameter vector $\bar{\theta}(k)$ corresponding to the imaginary plant $\bar{r}(z,k)\times\bar{p}(z,k)^{-1}$. It is to be noted that $\bar{\theta}(k)$ converges to the desired control parameter θ as $k\to\infty$. We, however, do not solve simultaneous equation (6.3) to obtain $\bar{\theta}(k)$, but use an adaptive method mentioed in 6.2.

6.4 Adaptive pole assignment. We have an imaginary plant

$$\bar{t}(z,k) = \bar{r}(z,k)\bar{p}(z,k)^{-1} \qquad (6.21)$$

at the instant k. The output of the imaginary plant due to the input u(k) is denoted by $\bar{y}(k)$. The pole assignment for the imaginary plant is described as follows. Let the polynomial equation be

$$\bar{k}(z,k)\bar{p}(z,k)+\bar{h}(z,k)\bar{r}(z,k)=q(z)[\bar{p}(z,k)-p_d(z)]. \qquad (6.22)$$

We denote the solution by

$$\bar{k}(z,k) = \bar{k}_{n-2}(k)z^{n-2}+ \cdots +\bar{k}_0(k), \qquad (6.23)$$

and

$$\bar{h}(z,k) = \bar{h}_{n-1}(k)z^{n-1}+ \cdots +\bar{h}_0. \qquad (6.24)$$

Now define

$$\bar{\theta}(k)=[-\bar{k}_{n-2}(k),\cdots,-\bar{k}_0(k),-\bar{h}_{n-1}(k),\cdots,-\bar{h}_0(k)]^T, \qquad (6.25)$$

and

$$\omega(k)=[\frac{z^{n-2}}{q(z)}u(k),\cdots,\frac{1}{q(z)}u(k),\frac{z^{n-1}}{q(z)}\bar{y}(k),\cdots,$$

$$\frac{1}{q(z)}\bar{y}(k)]^T. \tag{6.26}$$

In order to avoid solving simultaneous equation, we introduce $\hat{\theta}(k)$ which is the estimate of $\bar{\theta}(k)$ and is defined later. That is, instead of the control law

$$u(k) = -\bar{\theta}(k)^T\omega(k) + v(k), \tag{6.27}$$

we use the control law

$$u(k) = -\hat{\theta}(k)^T\omega(k) + v(k). \tag{6.28}$$

Now, we obtain from eq.(6.22) the relation

$$\bar{y}(k) = \frac{\bar{r}(z,k)}{p_d(z)}u(k) + \bar{\theta}(k)^T \frac{\bar{r}(z,k)}{p_d(z)}\omega(k), \tag{6.29}$$

which is the dynamics of the imaginary plant in terms of $\bar{r}(z,k)$, $p_d(z)$ and $\bar{\theta}(k)$. The identifier will then be defined by introducing the estimate $\hat{\theta}(k)$; i.e.,

$$\hat{y}(k) = \frac{\bar{r}(z,k)}{p_d(z)}u(k) + \hat{\theta}(k-1)^T \frac{\bar{r}(z,k)}{p_d(z)}\omega(k). \tag{6.30}$$

The identification error $\varepsilon_q(k)=\hat{y}(k)-\bar{y}(k)$ can be written as

$$\varepsilon_q(k) = [\hat{\theta}(k-1)-\bar{\theta}(k)]^T \frac{\bar{r}(z,k)}{p_d(z)}\omega(k). \tag{6.31}$$

We can use the following adaptive law, intending to make $\hat{\theta}(k)$ approach to $\bar{\theta}(k)$.

$$\hat{\theta}(k)=\hat{\theta}(k-1) - \frac{\bar{r}(z,k)p_d^{-1}(z)\omega(k)\varepsilon_q(k)}{c_q+[\bar{r}(z,k)p_d^{-1}(z)\omega(k)]^T[\bar{r}(z,k)p_d^{-1}(z,k)\omega(k)]},$$

$$c_q>0 \tag{6.32}$$

It is to be noticed that the unknown parameter vector $\bar{\theta}(k)$ is not constant but varies with time. However, it is further noticed that $\bar{\theta}(k)$ converges to a constant θ as $k\to\infty$.

6.5 Stability analysis. Consider a function

$$V(k) = [\hat{\theta}(k)-\bar{\theta}(k)]^T[\hat{\theta}(k)-\bar{\theta}(k)]. \tag{6.33}$$

For the simplicity, we define

$$\Omega(k) = \bar{r}(z,k)p_d^{-1}(z)\omega(k). \tag{6.34}$$

Then, after some calculation, we obtain

$$\Delta V(k) = ||\hat{\theta}(k-1)-\bar{\theta}(k)||^2 - ||\hat{\theta}(k-1)-\bar{\theta}(k-1)||^2 - \frac{2c_q + \Omega^T(k)\Omega(k)}{[c_q + \Omega^T(k)\Omega(k)]^2}\varepsilon_q^2(k). \tag{6.35}$$

From the Schwartz inequality, we obtain

$$||\hat{\theta}(k-1)-\bar{\theta}(k)||^2 - ||\hat{\theta}(k-1)-\bar{\theta}(k-1)||^2 \leq ||\bar{\theta}(k)-\bar{\theta}(k-1)||^2. \tag{6.36}$$

Therefore,

$$\Delta V(k) \leq ||\bar{\theta}(k)-\bar{\theta}(k-1)||^2 - \frac{2c_q+\Omega^T(k)\Omega(k)}{[c_q+\Omega^T(k)\Omega(k)]^2}\varepsilon_q^2(k). \tag{6.37}$$

However, since $\bar{\theta}(k)-\bar{\theta}(k-1)\rightarrow\infty$ as $\rightarrow\infty$, there exists K such that, for k>K the following holds.

$$\Delta V(k) \leq - \frac{2c_q+\Omega^T(k)\Omega(k)}{[c_q+\Omega^T(k)\Omega(k)]^2}\varepsilon_q^2(k). \tag{6.38}$$

The net results are such that $V(k)\rightarrow$const., $\Delta V(k)\rightarrow 0$, and $\varepsilon_q(k)\rightarrow 0$. On the assumption of persitent exciting, we obtain $\hat{\theta}(k)$ $\rightarrow\bar{\theta}(k)$ from eq.(6.31), which in turn implies $\hat{\theta}(k)\rightarrow\theta$. Thus the pole assignment is achieved.

It is to be noted that total number of parameters to be estimated is 4n-1. Simulation result is shown below. The plant transfer function is given as

$$t(z) = \frac{2(z+1.2)}{(z-0.5)(z+1.01)},$$

and the desired closed loop characteristic polynomial is chozen as z^2. In the simulation, q(z) is choze as z+0.2, and the reference input is taken as sgn (sin 0.8k). Initial values for $\tilde{r}(k)$ and $\tilde{p}(k)$ are both taken as $[1 \quad 1]^T$, while initial values for $\hat{\theta}(k)$ are taken as 0. Adaptive law based on least square is commonly used for both plant identification and updating of control parameters. The simulation result is shown in Fig.6.1.

REFERENCES

[1] H.Elliott and W.A.Wolovich;Parameter adaptive identification and control,IEEE Trans.Vol.AC-24,592/599,1979.

[2] K.J.Astrom;Direct methods for nonminimum phase systems, Proc. 19th Decision and Control,611/615,1980.

[3] G.Kreisselmeier;Adaptive control via adaptive observer and asymptotic feedback matrix synthesis,IEEE Trans. Vol.AC-25,717/722,1980.

[4] G.C.Goodwin and K.Sin; Adaptive control of nonminimum phase systems,IEEE Trans.Vol.AC-26,478/483,1981.

[5] H.Elliott; Direct pole placement with application to nonminimum phase systems,IEEE Trans. Vol.AC-27,720/722, 1982.

[6] W.A.Wolovich; Linear Multivariable Systems, Springer-Verlag,New York,1974.

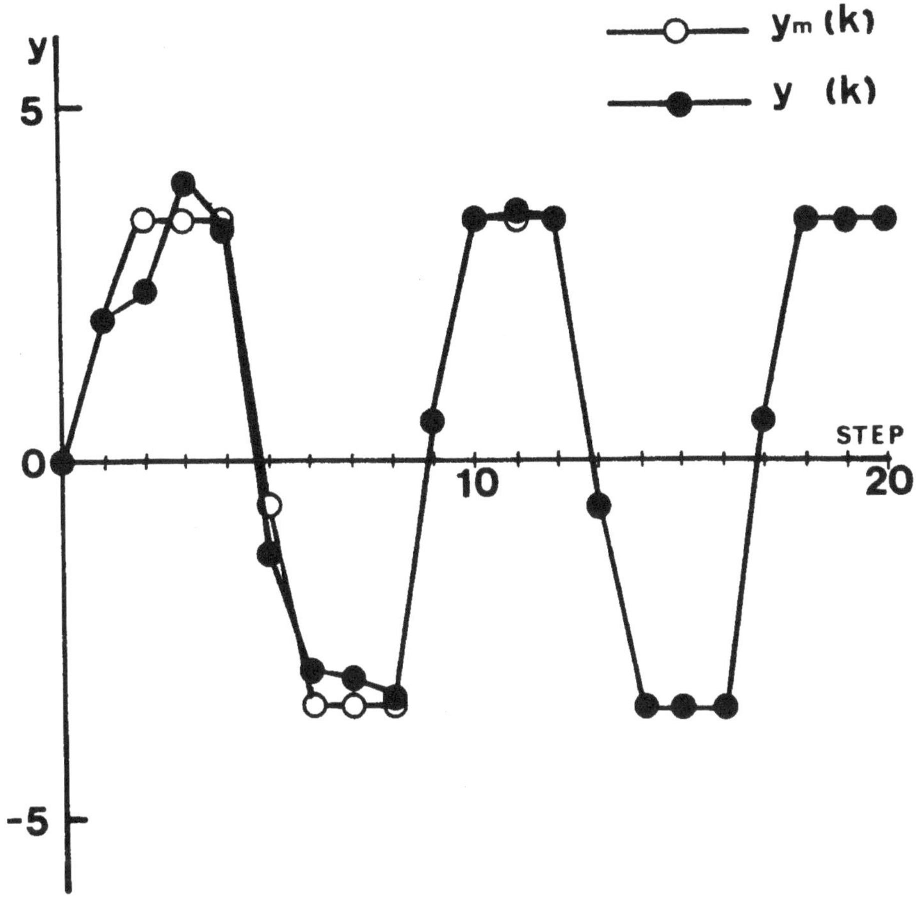

Fig. 6.1 Simulation of adaptive pole assignment

Chapter 7 Decoupling control with exact model matching

It is needless to say that the ideal feature of controlling multivariable systems is to control each output independently. This is called decoupling control. There are some who believe that the decoupling control is a very difficult problem or is unrealizable unless the plant to be controlled satisfies some strict conditions. All that is not true, because the decoupling control is easily solved as a natural extension of scalar exact model matching techniques explained in Chapter 3 and only the the very week condition that the plant is right invertible is needed for decoupling multivariable systems.

The extension of classical frequency response method praized in Great Britain [1] is an approximate method and does not suit to this lecture. Most of works on decoupling have been done in time domain [2]-[6], and do not present so simple guiding principle for design inspite of rather intricate arguments. There have been few researches whch have been done in frequency domain [7],[8].

The multivariable system is regarded as a set of p multi-input single-output systems in this lecture in order to extend exact model matching techniques to the decoupling control in a natural way. Such an idea was employed by Idegawa et al. in the design method of model following [9] but it was used in time domain method. The design method presented in this lecture not only achieves decoupling but also carries out exact model matching to each decoupled scalar system.

7.1 Description of the problem. Let the dynamics of m-input p-output plant be given by the transfer matrix

$$T(s) = [\bar{r}_{ij}(s)/p_{ij}(s)]. \qquad (7.1)$$

The reference model should be given by p×p diagonal matrix because the objective of control is decoupling, and then is denoted by

$$T_d(s) = \mathrm{diag}[g_{di}r_{di}(s)/p_{di}(s)], \qquad (7.2)$$

where $g_{di}r_{di}(s)/p_{di}(s)$ means a desired transfer function from reference input $v_i(s)$ to output $y_i(s)$. In the scalar ex-act model matching, an assumption on relative degrees was needed as well as an assumption that the plant is of non-minimum phase. In the case of multivariable systems, corresponding assumptions with above are also needed, but the first assumption required especially in multivariable case is

A.1 $\rho[T(s)] = p$.

This assumption can be rephrased that T(s) is right invertible. Clearly, $\rho[T(s)]=p$ requires $m \geq p$. The assumption A.1 is never an unreasonable one, because $\rho[T(s)]= p$ merely implies that each output is independent and nontrivial and is always satified in the plant with engineering significance. The assumption on relative degrees and of minimum phase cannot be stated so simply as for scalar systems and will be stated in the appropriate subsequent portions.

By reducing each row of plant transfer matrix to a common denominator, T(s) can be rewritten as

$$T(s) = \begin{bmatrix} \frac{g_{11}r_{11}(s)}{p_1(s)} & \cdots\cdots & \frac{g_{1m}r_{1m}(s)}{p_1(s)} \\ \vdots & & \vdots \\ \frac{g_{p1}r_{p1}(s)}{p_p(s)} & \cdots\cdots & \frac{g_{pm}r_{pm}(s)}{p_p(s)} \end{bmatrix}$$

$$= \begin{bmatrix} p_1(s) & & 0 \\ & \ddots & \\ 0 & & p_p(s) \end{bmatrix}^{-1} \begin{bmatrix} g_{11}r_{11}(s) & \cdots & g_{1m}r_{1m}(s) \\ \vdots & & \vdots \\ g_{p1}r_{p1}(s) & \cdots & g_{pm}r_{pm}(s) \end{bmatrix}$$

$$\underline{\underline{d}}\; P^{-1}(s)R(s), \tag{7.3}$$

where $r_{ij}(s)$'s and p_i's are monic. When ij element of $T(s)$ is zero, $r_{ij}(s)=1$ and $g_{ij}=0$. A numerical matrix called B^* plays an extremely important role, whatever approaches are employed for decoupling. In the first, representation of B^* by the parameters used in the state space is given. Let the state space representation of the plant be given by

$$\dot{x}(t) = Ax(t) + Bu(t) \tag{7.4a}$$

$$y(t) = Cx(t). \tag{7.4b}$$

The transfer matrix will be

$$\begin{aligned} T(s) &= C(sI-A)^{-1}B \\ &= \frac{1}{|sI-A|} C \operatorname{adj}(sI-A) B \\ &= \frac{1}{|sI-A|} C(Is^{n-1}+(A+a_{n-1}I)s^{n-2}+ \cdots \\ &\qquad +(A^{n-1}+a_{n-1}A^{n-2}+ \cdots +a_1 I))B. \end{aligned} \tag{7.5}$$

The $T_i(s)$, the i-th row of $T(s)$, will then be

$$\begin{aligned} T_i(s) &= \frac{1}{|sI-A|} \{C_iBs^{n-1}+(C_iAB+a_{n-1}C_iB)s^{n-2}+ \cdots \\ &\qquad +(C_iA^{n-1}B+a_{n-1}C_iA^{n-2}B+ \cdots +a_1C_iB)\}. \end{aligned} \tag{7.6}$$

Suppose that $C_iB=0$, $C_iAB=0$, $\cdots$, $C_iA^{d_i-1}B=0$, but $C_iA^{d_i}B\neq 0$. Then, the i-th row degree of numerator polynomial matrix of T(s) is evaluated as $n-d_i-1$, and the row vector con-sisting of coefficients of $n-d_i-1$ degree terms must be $C_iA^{d_i}B$. The matrix B^* is now defined as

$$B^* = \begin{bmatrix} C_1A^{d_1}B \\ \vdots \\ C_pA^{d_p}B \end{bmatrix}. \tag{7.7}$$

That is to say, B^* is $\Gamma_r[C \operatorname{adj}(sI-A)\, B]$.

As is seen, the index d_i determined for each row is defined as the least integer d_j such that $C_iA^{d_j}B$ is nonzero, but it can also be defined as the integer d_i such that

$$\lim_{s\to\infty} s^{d_i+1}T_i(s) = \text{nonzero finite vector} \tag{7.8}$$

is satisfied, where the nonzero finite row vector is nothing but $C_iA^{d_i}B$. By using d_i defined as above, B^* can also be written as

$$B^* = \lim_{t\to\infty} \begin{bmatrix} s^{d_1+1} & & 0 \\ & \ddots & \\ 0 & & s^{d_p+1} \end{bmatrix} T(s). \tag{7.9}$$

Further, let $\partial[p_i(s)]=n_i$ and $\max_j D\partial[r_{ij}(s)]=m_i$, then since $d_i+1=n_i-m_i$, another representation for B^* is obtained as

$$B^* = \lim_{t\to\infty} \begin{bmatrix} s^{n_1-m_1} & & 0 \\ & \ddots & \\ 0 & & s^{n_p-m_p} \end{bmatrix} T(s). \tag{7.10}$$

Let us define sets J_i, $i=1,2,\cdots,p$ by

$$J_i = \{j \mid \partial[r_{ij}(s)]=m_i\}. \tag{7.11}$$

Then it follows that

$$B^* = \{B^*_{ij}\},\ B^*_{ij} = \begin{cases} g_{ij} & \text{if } j\in J_i \\ 0 & \text{if } j\notin J_i \end{cases}. \tag{7.12}$$

Finally, the more brief representation is

$$B^* = \Gamma_r[R(s)]. \tag{7.13}$$

Since $\rho[T(s)]=p$, $\rho[R(s)]=p$, i.e., $R(s)$ has full row rank, but $\rho[B^*]$ may or may not have full row rank p. Also m may be equal to or greater than p. Then, we will classify the decoupling problem into four kinds of problem P1, P2, P3, and P4 as shown in Table 7.1. For P1 and P2, $\rho[R(s)]=p$ and $\rho[B^*]=p$, whereas for P3 and P4 $\rho[R(s)]=p$ but $\rho[B^*]<p$. As will be mentioned later, P3 and P4 can be reduced to P1 and P2 respectively by emplying a dynamic compensator. Also P2 can always be reduced to P1. Therefore all kinds of decoupling problem are essentially reduced to P1.

We are now in the situation to describe the assumption on relative degrees.

A.2 $n_{d_i} - m_{d_i} \geq n_i - m_i$, $i=1,2,\cdots,p$, where $n_{d_i}=\partial[p_{d_i}(s)]$ and $m_{d_i}=\partial[r_{d_i}(s)]$.

Introduce m_i degree $r^*_i(s)$ and n_i degree $p^*_i(s)$, each being any monic stable polynomial, $i=1,2,\cdots,p$. Then, $T_d(s)$ can be rewritten as follows.

$$T_d(s) = \begin{bmatrix} \frac{g_{d1}r^*_1(s)}{p^*_1(s)} & & 0 \\ & \ddots & \\ 0 & & \frac{g_{dp}r^*_p(s)}{p^*_p(s)} \end{bmatrix} \begin{bmatrix} \frac{r_{d1}(s)\,p^*_1(s)}{r^*_1(s)p_{d1}(s)} & & 0 \\ & \ddots & \\ 0 & & \frac{r_{dp}(s)p^*_p(s)}{r^*_p(s)p_{dp}(s)} \end{bmatrix}.$$

Table 7.1 Classification of decoupling problem

	m = p	m > p
$\rho[B^*] = p$	P 1	P 2
$\rho[B^*] < p$	P 3	P 4

(7.14)

Since diag $[r_{di}(s)p^*_i(s)/r^*_i(s)p_{di}(s)]$ is proper and stable, it can be taken as an input dynamics $T_{IN}(s)$. Denote $T_{IN}(s)v(s)$ by $\bar{v}(s)$, and $\mathrm{diag}[g_{di}r^*_i(s)/p^*_i(s)]$ by $T^*(s)$.

7.2 Dynamic compensation. In problems P3 and P4, R(s) is of full row rank, but is not row proper. It is known that R(s) is column equivalent to a lower left triangular matrix [9]; i.e., one can always find a unimodular matrix $U_R(s)$ such that $R(s)U_R(s)$ is a lower left triangular matrix. Since R(s) is of full row rank, all diagonal elements of $R(s)U_R(s)$ are nonzero and hence $R(s)U_R(s)$ must be row proper. Since $U_R(s)$ is a differential operator, it is not usuable for precompensator. Then, take any monic polynomial d(s), the degree of which is equal to $\partial[U_R(s)]$ and consider a system described by $U_R(s)/d(s)$. Since $U_R(s)/d(s)$ is proper and hence realizable, it can be employed as a dynamic compensator to the plant T(s). Regard the plant with serial connection of dynamic compensator as a new augmented plant. The transfer matrix of the augmented plant will be

$$\begin{aligned} &T(s)U_R(s)/d(s) \\ &= P^{-1}(s)R(s)U_R(s)/d(s) \\ &= [P(s)d(s)]^{-1}[R(s)U_R(s)], \end{aligned} \tag{5.15}$$

where $R(s)U_R(s)$ is of full row rank and row proper. Thus, P3 and P4 are reduced to P1 and P2 respectively. Further, as will be explained later, P2 can be easily solved once the solving method for P1 is revealed. Therefore, it suffices to concentrate ourselves to the solving method for P1.

7.3 Solution of P1 problem. In P1, $m=p$, $T(s)$ is nonsingular, and $B^*=\Gamma_r[R(s)]$ is nonsingular too. We can now state on the third assumption.

A.3 $R(s)$ is asymptotically stable.

Let us insert a statical element $(B^*)^{-1}$ before the plant, and regard $T(s)(B^*)^{-1}$ as a new plant;

$$T(s)(B^*)^{-1} = P^{-1}(s)R(s)\Gamma_r^{-1}[R(s)]$$
$$\stackrel{d}{=} P^{-1}(s)R^B(s), \tag{7.16}$$

where $R^B(s)$ has the same column degrees as $R(s)$ and $\Gamma_r[R^B(s)]=I$. Notice that if we set $R^B(s)=[g^B_{ij}r^B_{ij}(s)]$, $\partial[r^B_{ii}(s)]=m_i$, $\partial[r^B_{ij}(s)]\leq m_i$ for $i\neq j$, and $g^B_{ii}=1$. Denote the input signal to the statical element $(B^*)^{-1}$ by $u^B(s)$.

We want to apply the scalar exact model matching techniques to this decoupling problem. For each element of the i-th row, polynomials $k(s)$ and $h(s)$ should be determined. We consider first the i-th element, i.e., the diagonal element $g^B_{ii}r^B_{ii}(s)$, where $g^B_{ii}=1$ and $\partial[r^B_{ii}(s)]=m_i$. Introduce any n_i-m_i-1 degree monic stable polynomial $\tau_i(s)$, and set up the equation

$$k^B_{ii}(s)p_i(s)+h^B_{ii}(s)r^B_{ii}(s) = \tau_i(s)r^*_i(s)p_i(s)$$
$$-\tau_i(s)r^B_{ii}(s)p^*_i(s). \tag{7.17}$$

It is to be noticed that $p_i(s)$ and $r^B_{ii}(s)$ are not necessarily relatively prime unlike the case for the scalar problem. Suppose first that $p_i(s)$ and $r^B_{ii}(s)$ are relatively prime. From eq.(7.17), we obtain

$$k^B_{ii}(s)-\tau_i(s)r^*(s) = -\frac{r^B_{ii}(s)}{p_i(s)}\{h^B_{ii}(s)+\tau_i(s)p^*_i(s)\}$$

(7.18)

Since $p_i(s)$ and $r^B_{ii}(s)$ are relatively prime and the right hand side of eq.(7.18) must reduce to a polynomial, $2n_i-m_i-1$ degree $h^B_{ii}(s)+\tau_i(s)p^*_i(s)$ must be divisible by n degree $p_i(s)$. That is, if we set the quotient and residual of $\tau_i(s)p^*_i(s)/p_i(s)$ as $\alpha_i(s)$ of n_i-m_i-1 degree and $r_{e,i}(s)$ of n_i-1 degree respectively, $h^B_{ii}(s)$ must be identical with $-r_{e,i}(s)$, which depends only on $\tau_i(s)$, $p^*_i(s)$ and $p_i(s)$ without any relation to $r^B_{ii}(s)$. Then, we denote $h^B_{ii}(s)$ by $h^B_i(s)$. The polynomial $k^B_{ii}(s)$ is determined as $\tau_i(s)r^*_i(s)-r^B_{ii}(s)\alpha_i(s)$ and is of degree n_i-2.

Suppose next that $r^B_{ii}(s)$ and $p_i(s)$ are not relatively prime. Then the polynomials $k^B_{ii}(s)$ and $h^B_{ii}(s)$ which satisfy eq.(7.18) are not unique, but it is justified to use $h^B_i(s)$ as $h^B_{ii}(s)$ because then $k^B_{ii}(s)$ is still determined as $\tau_i(s)r^*_i(s)-r^B_{ii}(s)\alpha_i(s)$. That is, $h^B_i(s)$ can always be taken as $h^B_i(s)$.

Secondly, let us consider the elements other than the i-th element in the i-th row. Then $g^B_{ij}r^B_{ij}(s)$ is of degree lower than m_i, and hence $\tau_i(s)g^B_{ij}r^B_{ij}(s)p^*_i(s)$ is of degree $2n_i-2$ at most. Therefore, we can set up the equation as

$$k^B_{ij}(s)p_i(s) + h^B_{ij}(s)g^B_{ij}r^B_{ij}(s) = -\tau_i(s)g^B_{ij}r^B_{ij}(s)p^*_i(s). \qquad (7.19)$$

Again, it is justified to use $h^B{}_i(s)$ as $h^B{}_{ij}(s)$ because then $k^B{}_{ij}(s)$ is determined as $-g^B{}_{ij}r^B{}_{ij}(s)\alpha_i(s)$ of degree n_i-2 at most. As explained above, $h^B{}_{ij}(s)$, $j=1,2,\cdots,p$ $(=m)$, can always be taken as n_i-1 degree $h^B{}_i(s)$ regardless of j. This fact is stated as a lemma.

Lemma 7.1 (polynomial $h^B{}_i(s)$) Polynomial $h^B{}_{ij}(s)$ which satisfies the following equations can be taken as $-\gamma_i(s)p^*{}_i(s)$ mod $p_i(s)$ $(=-r_{e,i}(s))$ regardless of j, which is denoted by $h^B{}_i(s)$.

for $j=i$,

$$k^B{}_{ii}(s)p_i(s)+h^B{}_{ii}(s)r^B{}_{ii}(s) = \gamma_i(s)r^*{}_i(s)p_i(s)$$
$$- \gamma_i(s)r^B{}_{ii}(s)p^*{}_i(s)$$

and for $j\neq i$,

$$k^B{}_{ij}(s)p_i(s)+h^B{}_{ij}(s)g^B{}_{ij}r^B{}_{ij}(s)$$
$$= -\gamma_i(s)g^B{}_{ij}r^B{}_{ij}(s)p^*{}_i(s).$$

Furthermore, $h^B{}_i(s)$ is of degree n_i-1 while $k^B{}_{ij}(s)$ is of degree n_i-2.

Theorem 7.1 (control law). The control law

$$u^B{}_i(s) = \sum_{j=1}^{p} k^B{}_{ij}(s)[\gamma_i(s)r^*{}_i(s)]^{-1}\cdot u^B{}_j(s)$$
$$+ h^B{}_i(s)[\gamma_i(s)r^*{}_i(s)]^{-1}\cdot y_i(s) + g_{di}\bar{v}_i(s),$$
$$i=1,2,\cdots,p \tag{7.20}$$

is feasible and achieves decoupling with each decoupled scalar system being exactly model matched.

(proof) Since all $k^B_{ij}(s)[\tau_i(s)r^*_i(s)]^{-1}$ and $h^B_i(s)[\tau_i(s)r^*(s)]^{-1}$ are proper, the control law is feasible. When $j=i$, we have eq.(7.17), while we have eq.(7.19) when $j \neq i$. From these equations, we obtain

$$k^B_{ii}(s)[\tau_i(s)r^*_i(s)]^{-1} \cdot u^B_i(s)$$
$$+ h^B_i(s)[\tau_i(s)r^*_i(s)]^{-1} \cdot r^B_{ii}(s)p_i^{-1}(s) \cdot u^B_i(s)$$
$$= u^B_i(s) - p^*_i(s)[r^*(s)]^{-1} \cdot r^B_{ii}(s)p_i^{-1}(s) \cdot u^B_i(s),$$

and

$$k^B_{ij}(s)[\tau_i(s)r^*_i(s)]^{-1} \cdot u^B_j(s)$$
$$+ h^B_i(s)[\tau_i(s)r^*_i(s)]^{-1} \cdot g^B_{ij}r^B_{ij}(s)p_i^{-1}(s) \cdot u^B_j(s)$$
$$= -p^*_i(s)[r^*_i(s)]^{-1} \cdot g^B_{ij}r^B_{ij}(s)p_i^{-1}(s) \cdot u^B_j(s).$$

Summing up these equations for all j, we obtain

$$\sum_{j=1}^{p} k^B_{ij}(s)[\tau_i(s)r^*_i(s)]^{-1} \cdot u^B_{j}(s)$$
$$+ h^B_i(s)[\tau_i(s)r^*_i(s)]^{-1} \cdot y_i(s)$$
$$= u^B_i(s) - p^*_i(s)[r^*_i(s)]^{-1} \cdot y_i(s).$$

Then, by the control law (7.20), it results that

$$p^*_i(s)[r^*_i(s)]^{-1} \cdot y_i(s) = g_{di}\bar{v}_i(s),$$

i.e.,

$$y_i(s) = \frac{g_{di}r^*_i(s)}{p^*_i(s)} \bar{v}_i(s).$$

Thus, finally it results

$$y(s) = T^*(s)\bar{v}(s).$$

Further, we obtain

$$u(s) = R^{-1}(s)P(s)T^*(s)\bar{v}(s).$$

Since R(s) is assumed to be stable, u(t) must be bounded for bounded $\bar{v}(t)$. Q.E.D.

The vector form of control law (7.20) is presented in the following. Let

$$\left.\begin{aligned} u^B(s) &= [u^B{}_1(s) \cdots u^B{}_p(s)]^T \\ y(s) &= [y_1(s) \cdots y_p(s)]^T \\ \Gamma(s)R^*(s) &= \text{diag}\ [\gamma_1(s)r^*{}_1(s) \cdots \gamma_p(s)r^*{}_p(s)] \\ K^B(s) &= [k^B{}_{ij}(s)] \\ H^B(s) &= \text{diag}\ [h^B{}_1(s) \cdots h^B{}_p(s)] \\ G_d &= \text{diag}\ [g_{d1} \cdots g_{dp}] \\ \bar{v}(s) &= [\bar{v}_1(s) \cdots \bar{v}_p(s)]^T, \end{aligned}\right\} \quad (7.21)$$

then eq.(7.20) can be rewritten as

$$u^B(s) = [\Gamma(s)R^*(s)]^{-1}K^B(s)u^B(s) + [\Gamma(s)R^*(s)]^{-1}H^B(s)y(s) + G_d\bar{v}(s). \quad (7.22)$$

Furthermore, by using the relation $u(s)=[B^*]^{-1}u^B(s)$, the control law expressed in terms of u(s) becomes

$$u(s) = [B^*]^{-1}\{[\Gamma(s)R^*(s)]^{-1}K(s)u(s) + [\Gamma(s)R^*(s)]^{-1}H(s)y(s)\} + [B^*]^{-1}G_d\bar{v}(s), \quad (7.23)$$

where

$$K(s) = K^B(s)B^*,\ H(s) = H^B(s). \quad (7.24)$$

The decoupling + exact model matching system obtained by solving P1 problem is shown in Fig.7.1.

7.4 Solution of P2 problem. In P2, m>p and both R(s) and $B^*=\Gamma_r[R(s)]$ have full rank p. Then we can choose p indep-

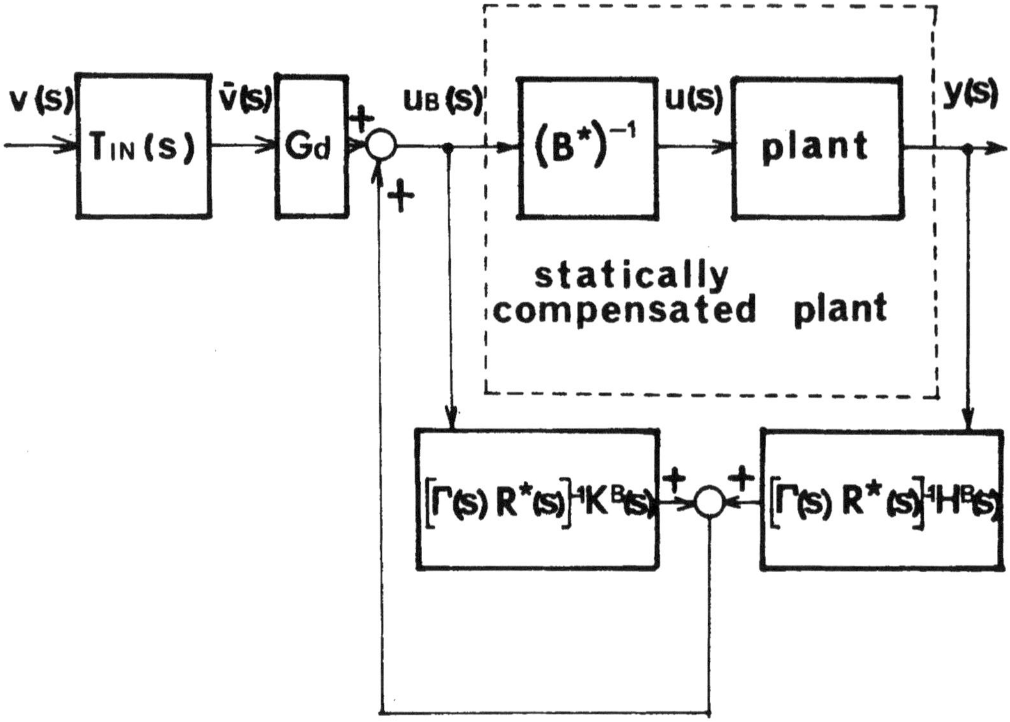

Fig. 7.1 Decoupling control system with exact model matching for P1 problem

endent columns among B^*. The way to choose such columns is not unique. Let the numbers of columns chosen be j_1, j_2, $\cdots$, j_p. These columns consitute a nonsingular matrix B^*_s. Corresponding to the above, let us take a square matrix $R_s(s)$ by picking up the columns with the same number from $R(s)$. Since B^*_s has no zero rows, the row degree of $R_s(s)$ is the same as $R(s)$; i.e., $\partial_{r_i}[R_s(s)]=m_i$, $i=1,2,\cdots,p$. Therefore $\Gamma_r[R_s(s)]=B^*_s$, which implies that $R_s(s)$ is nonsingular.

Let the components of $u(s)$ other then j_1, j_2, $\cdots$, j_p elements set to zero. Then it suffices to find

$$u_s(s) = [u_{j1}(s) \quad u_{j2}(s) \cdots u_{jp}(s)]^T. \tag{7.25}$$

Since

$$\begin{aligned} y(s) &= T(s)u(s) \\ &= P^{-1}(s)R(s)u(s) \\ &= P^{-1}(s)R_s(s)u_s(s), \end{aligned} \tag{7.26}$$

the problem P2 has been reduced to P1 by regarding the plant as $P^{-1}(s)R_s(s)$ and considering the problem to find $u_s(s)$. Once $u_s(s)$ has been determined following the solution method for P1, it only remains to insert 0 appropriately to constitute $u(s)$ such that j_1, j_2, $\cdots$, j_p elements of $u(s)$ be identical with 1, 2, $\cdots$, p elements of $u_s(s)$.

In order for the reduced problem to be solved, $R_s(s)$ must be stable as is indicated in A.3. This assumption will be restated as an assumption for P2 problem; i.e.,

A.3' There exist a stable $R_s(s)$.

7.5 Generic multivariable exact model matching. The generic exact model matching problem, where $T_d(s)$ is not diagonal nor nonsingular, seems to have little engineering meaning. Suppose the reference model transfer matrix of p×q form be given by

$$T_d(s) = P_d^{-1}(s)R_d(s), \tag{7.27}$$

where

$$P_d(s) = \begin{bmatrix} p_{d1}(s) & & 0 \\ & \ddots & \\ 0 & & p_{dp}(s) \end{bmatrix} \tag{7.28}$$

and

$$R_d(s) = \begin{bmatrix} g_{d11}r_{d11}(s) & \cdots & g_{d1q}r_{d1q}(s) \\ \vdots & & \vdots \\ g_{dp1}r_{dp1}(s) & \cdots & g_{dpq}r_{dpq}(s) \end{bmatrix}. \tag{7.29}$$

Also, let $\partial[p_{di}(s)] = n_{di}$ and $\max \partial[r_{dij}(s)] = m_{di}$.

Assume the assumption A.2 introduced in decoupling control again in this case. Introduce any n_i and m_i degree monic stable polynomials $p^*_i(s)$ and $r^*_i(s)$ respectively, i=1,2, ···,p. Define input dynamics by

$$T_{IN}(s) = \begin{bmatrix} \dfrac{p^*_1(s)}{p_{d1}(s)} \cdot \dfrac{[g_{d11}r_{d11}(s)\cdots g_{d1q}r_{d1q}(s)]}{r^*_1(s)} \\ \vdots \\ \dfrac{p^*_1(s)}{p_{dp}(s)} \cdot \dfrac{[g_{dp1}r_{dp1}(s)\cdots g_{dpq}r_{dpq}(s)]}{r^*\ (s)} \end{bmatrix} \tag{7.30}$$

Then factorize T(s) as follows.

$$T(s) = T^*(s)T_{IN}(s), \tag{7.31}$$

where

$$T^*(s) = \begin{bmatrix} \frac{r^*_1(s)}{p^*_1(s)} & & 0 \\ & \ddots & \\ 0 & & \frac{r^*_p(s)}{p^*_p(s)} \end{bmatrix}. \qquad (7.32)$$

Then, by putting $T_{IN}(s)v(s)$ as $\bar{v}(s)$, the problem is indeed reduced to the decoupling control problem. Therefore, the assumptions A.1 and A.3 (or A.3') are needed in addition to A.2.

REFERENCES

[1] M. Araki; CAD for multivariable systems: INA method and its recent development (1,2,3), System and Control 26,218/227,353/362,489/497 (in Japanese), 1982.

[2] P.L. Falb and Wolovich; Decoupling in the design and synthesis of multivariable control systems,IEEE Trans. Vol.AC-12,651/659,1967.

[3] E.G. Gilbert; The decoupling control of multivariable systems by state feedback, J. of SIAM Control,7,50/63, 1969.

[4] M. Kohno and I. Sugiura; Generalization of non-interacting control, SICE Trans.,10,589/593 (in Japanese), 1974.

[5] N. Kobayashi and T. Nakamizo; Decoupling and pole assignment,ibid,18,317/323(in Japanese), 1982.

[6] N. Kobayashi and T. Nakamizo; On stability of decoupling system with dynamic compensator,ibid, 19,97/103(in Japanese),1983.

[7] H. Hikita; On diagonal decoupling of linear systems with rectangular transfer function matrix,ibid,19,451/

457(in Japanese),1983.

[8] K. Ichikawa; Two-step decoupling algorithm for multivariable systems,Int. J. Control,38,1239/1247,1983.

[9] T. Degawa, K. Kanai, and S. Uchikado; On a method of designing a multivariable model following control system,SICE Trans.,18,1132/1139(in Japanese),1982.

Chapter 8 Multivariable adaptive control

We consider an adaptive control corresponding to P1 defined in the previous chapter; i.e., the problem with $m=p$ and with unknown but nonsingular B^*. It is to be noticed that quite the same as P1, three conditions A.1, A.2 and A.3 are sufficient and necessary for adaptive control to be achieved. There have been many researches on multivariable adaptive control [1]-[4], but the approach presented here is not so resemble to them. In partiular, the approach is an extension of the theory on decoupling with exact model matching mentioned in the previous chapter to the adaptive case.

8.1 Control law. The polynomial eqs.(7.17) and (7.19) can be written in matrix form; i.e.,

$$P(s)K^B(s) + H^B(s)R^B(s) = \Gamma(s)R^*(s)P(s) - \Gamma(s)P^*(s)R^B(s), \tag{8.1}$$

where $P(s)$ and $R^B(s)$ are defined eqs.(7.3) and (7.16) respectively, $K^B(s)$ and $H^B(s)$ are defined by eq.(7.21), and

$$\Gamma(s) = \mathrm{diag}[\tau_1(s) \cdots \tau_p(s)], \tag{8.2}$$

$$P^*(s) = \mathrm{diag}[p^*_1(s) \cdots p^*_p(s)]. \tag{8.3}$$

Using relations (7.16) and (7.24), eq.(8.1) can be transformed into

$$P(s)K(s) + H(s)R(s) = \Gamma(s)R^*(s)P(s)B^* - \Gamma(s)P^*(s)R(s). \tag{8.4}$$

On the other hand, the control law described by eq.(7.23) can be rewritten as

$$u(s) = (B^*)^{-1}\left\{\begin{bmatrix} \frac{1}{T_1(s)r^*_1(s)} & & \\ & \ddots & \\ & & \frac{1}{T_1(s)r^*_1(s)} \end{bmatrix}\right.$$

$$\times\begin{bmatrix} k_{11}(s) & \cdots & k_{1p}(s) \\ & \vdots & \\ k_{p1}(s) & \cdots & k_{pp}(s) \end{bmatrix} u(s) + \begin{bmatrix} \frac{1}{T_1(s)r^*_1(s)} & & \\ & \ddots & \\ & & \frac{1}{T_p(s)r^*_p(s)} \end{bmatrix}$$

$$\left.\times\; y(s)\right\} + (B^*)^{-1}G_d\,\bar{v}(s)$$

$$= (B^*)^{-1}\left\{\begin{matrix} \frac{1}{T_1(s)r^*_1(s)}[k_{11}(s)u_1(s) + \cdots + k_{1p}(s)u_p(s)] \\ \vdots \\ \frac{1}{T_p(s)r^*_p(s)}[k_{p1}(s)u_1(s) + \cdots + k_{pp}(s)u_p(s)] \\ + \frac{h_1(s)}{T_1(s)r^*_1(s)}\,y_1(s) \\ \\ + \frac{h_p(s)}{T_p(s)r^*_p(s)}\,y_p(s) \end{matrix}\right\} + (B^*)^{-1}G_d\,\bar{v}(s). \quad (8.5)$$

Let the polynomial $k_{ij}(s)$ be represented by

$$k_{ij}(s) = k^{ij}_{n-2}s^{n-2} + \cdots + k^{ij}_0,\ i,j=1,2,\cdots,p. \quad (8.6)$$

Also, let the polynomial $h_i(s)$ be represented by

$$h_i(s) = h^i_{n-1}s^{n-1} + \cdots + h^i_0,\quad i=1,2,\cdots,p. \quad (8.7)$$

Further, let us define parameter vectors θ_i, $i=1,2,\cdots,p$ by

$$\theta_i = [-k^{i1}_{n-2}, \cdots, -k^{i1}_0, \cdots, -k^{ip}_{n-2}, \cdots, -k^{ip}_0, -h^i_{n-1}, \cdots, -h^i_0],\ i=1,2,\cdots,p \quad (8.8)$$

and signal vectors $\omega_i(t)$, i=1,2,···,p, by

$$\omega_i(t) = \Big(\frac{p^{n-2}}{\gamma_i(p)r^*_i(p)}u_1(t), \cdots, \frac{1}{\gamma_i(p)r^*_i(p)}u_1(t),$$
$$\cdots\cdots\cdots\cdots\cdots\cdots$$
$$\frac{p^{n-2}}{\gamma_i(p)r^*_i(p)}u_p(t), \cdots, \frac{1}{\gamma_i(p)r^*_i(p)}u_p(t),$$
$$\frac{p^{n-1}}{\gamma_i(p)r^*_i(p)}y_i(t), \cdots, \frac{1}{\gamma_i(p)r^*_i(p)}y_i(t)\Big),$$
$$i=1,2, \cdots, p. \tag{8.9}$$

Then, eq.(8.5), the control law, can be written as

$$u(t) = -(B^*)^{-1}\begin{pmatrix}\theta_1{}^T\omega_1(t)\\ \vdots \\ \theta_p{}^T\omega_p(t)\end{pmatrix} + (B^*)^{-1}G_d\,\bar{v}(t), \tag{8.10}$$

where θ_i and $\omega_i(t)$ are of dimension $p(n-1)+n$.

In the adaptive control, B^* and θ_i, i=1,2,···,p are unknown. Then, we set the control law as

$$u(t) = -[\tilde{B}^*(t)]^{-1}\begin{pmatrix}\tilde{\theta}_1{}^T(t)\omega_1(t)\\ \vdots \\ \tilde{\theta}_p{}^T(t)\omega_p(t)\end{pmatrix} + [\tilde{B}^*(t)]^{-1}G_d\bar{v}(t), \tag{8.11}$$

where $\tilde{B}^*(t)$ and $\tilde{\theta}_i(t)$, i=1,2,···,p, are the estimates of B^* and θ_i, i=1,2,···,p respectively and will be defined later.

8.2 Error dynamics. We need plant dynamics represented by B^*, $K(s)$ and $H(s)$ instead of $R(s)$ and $P(s)$. Multiplying eq.(8.4) by $[P(s)\Gamma(s)P^*(s)]^{-1}$ from the left, we obtain

$$[\Gamma(s)P^*(s)]^{-1}K(s) + [\Gamma(s)P^*(s)]^{-1}H(s)P^{-1}(s)R(s)$$
$$= [P^*(s)]^{-1}R^*(s)B^* - P^{-1}(s)R(s), \tag{8.12}$$

where the fact that $P(s)$, $\Gamma(s)$ and $P^*(s)$ are all diagonal is utilized. Multiplying u(s) from the left, and using the relation $y(s)=P^{-1}(s)R(s)u(s)$, we obtain

$$[\Gamma(s)P^*(s)]^{-1}K(s)u(s) + [\Gamma(s)P^*(s)]^{-1}H(s)y(s)$$
$$= (P^*(s)]^{-1}R^*(s)B^*u(s) - y(s),$$

or

$$y(s) = [P^*(s)]^{-1}R^*(s)B^*u(s) - [\Gamma(s)P^*(s)]^{-1}K(s)u(s) - [\Gamma(s)P^*(s)]^{-1}H(s)y(s), \quad (8.13)$$

which is further rewritten as

$$y(s) = [P^*(s)]^{-1}R^*(s)B^*u(s) - [P^*(s)]^{-1}R^*(s) \times\{[\Gamma(s)R^*(s)]^{-1}K(s)u(s)+[\Gamma(s)R^*(s)]^{-1}H(s)y(s)\}. \quad (8.14)$$

Using notations θ_i and $\omega_i(t)$, eq.(8.14) can be written as

$$y(t) = \begin{bmatrix} b^*{}_1{}^T \dfrac{r^*{}_1(p)}{p^*{}_1(p)}u(t) \\ \vdots \\ b^*{}_p{}^T \dfrac{r^*{}_p(p)}{p^*{}_p(p)}u(t) \end{bmatrix} + \begin{bmatrix} \theta_1{}^T \dfrac{r^*{}_1(p)}{p^*{}_1(p)}\omega_1(p) \\ \vdots \\ \theta_p{}^T \dfrac{r^*{}_p(p)}{p^*{}_p(p)}\omega_p(t) \end{bmatrix}, \quad (8.15)$$

where $b^*{}_i{}^T$ is the ith row of B^*. Eq.(8.15) is the desired representation of the plant dynamics.

The adaptive error e(t) is a p dimensional vector defined by $e(t)=y(t)-y_m(t)$. That is, the error dynamics is represented by

$$e(t) = \begin{bmatrix} b^*{}_1{}^T \dfrac{r^*{}_1(p)}{p^*{}_1(p)} \\ \vdots \\ b^*{}_p{}^T \dfrac{r^*{}_p(p)}{p^*{}_p(p)}u(t) \end{bmatrix} + \begin{bmatrix} \theta_1{}^T \dfrac{r^*{}_1(p)}{p^*{}_1(p)}\omega_1(t) \\ \vdots \\ \theta_p{}^T \dfrac{r^*{}_p(p)}{p^*{}_p(p)}\omega_p(t) \end{bmatrix} - y_m(t). \quad (8.16)$$

8.3 Adaptive law. The identifier for the error dynamics (8.16) is defined by

$$\tilde{e}(t) = \begin{bmatrix} [\tilde{b}^*_1(t)]^T \dfrac{r^*_1(p)}{p^*_1(p)}u(t) \\ \vdots \\ [\tilde{b}^*_p(t)]^T \dfrac{r^*_p(p)}{p^*_p(p)}u(t) \end{bmatrix} + \begin{bmatrix} \tilde{\theta}_1^T(t) \dfrac{r^*_1(p)}{p^*_1(p)}\omega_1(t) \\ \vdots \\ \tilde{\theta}_p^T(t) \dfrac{r^*_p(p)}{p^*_p(p)}\omega_p(t) \end{bmatrix}$$

$$- y_m(t). \tag{8.17}$$

The identification error $\varepsilon(t)$ is $\tilde{e}(t)-e(t)$; i.e.,

$$\varepsilon(t)=\begin{bmatrix} [\tilde{b}^*_1(t)-b^*_1]^T \dfrac{r^*_1(p)}{p^*_1(p)}u(t)+[\tilde{\theta}_1(t)-\theta_1]^T \dfrac{r^*_1(p)}{p^*_p(p)}\omega_1(t) \\ \vdots \\ [\tilde{b}^*_p(t)-b^*_p]^T \dfrac{r^*_p(p)}{p^*_p(p)}u(t)+[\tilde{\theta}_p(t)-\theta_p]^T \dfrac{r^*_p(p)}{p^*_p(p)}\omega_p(t) \end{bmatrix} \tag{8.18}$$

Let

$$\Omega_i(t) = [\frac{r^*_i(p)}{p^*_i(p)}u^T(t) \quad \frac{r^*_i(p)}{p^*_i(p)}\omega_i^T(t)]^T, \quad i=1,2,\cdots,p. \tag{8.19}$$

Then, eq.(8.18) can be written as

$$\varepsilon_i(t)=\left([\tilde{b}^*_i(t)-b^*_i]^T \quad [\tilde{\theta}_i(t)-\theta_i]^T\right)\Omega_i(t), \quad i=1,2,\cdots,p. \tag{8.20}$$

The following is an adequate adaptive law.

$$\left.\begin{aligned} \begin{bmatrix} \tilde{b}^*_i(t) \\ \tilde{\theta}_i(t) \end{bmatrix}^{\cdot} &= - \Gamma_i \frac{\Omega_i(t)\ \varepsilon_i(t)}{c + \Omega_i^T(t)\Omega_i(t)} \\ \Gamma_i=\Gamma_i^T>0, &\quad c_i>0 \end{aligned}\right\} i=1,2,\cdots,p. \tag{8.21}$$

The stability analysis under the adaptive law (8.21) is quite the same as was mentioned in Chapter 4. That is, in the first, we obtain

$$\lim_{t\to\infty} \frac{\varepsilon_i^{\,2}(t)}{c_i + \Omega_i^T(t)\Omega_i(t)} = 0. \qquad (8.22)$$

On the assumption that $||\Omega_i(t)||$ is bounded eq.(8.22) yields $\lim_{t\to\infty} \varepsilon_i(t)=0$, and hence $\lim_{t\to\infty} [\tilde{b}^*_i(t)]^{\cdot}=0$ and $\lim_{t\to\infty} [\tilde{\vartheta}_i(t)]^{\cdot}=0$ are obtained from eq.(8.21). The boundedness of $||\Omega_i(t)||$, however, is not yet assured. Suppose that $||\Omega_i(t)||$ is not bounded. Then, from eq.(8.22), the diverging velocity of $|\varepsilon_i(t)|$, if it would diverge, must be of lower order than $||\Omega_i(t)||$. Hence, again from eq.(8.21), we obtain $\lim_{t\to\infty} [\tilde{b}^*_i(t)]^{\cdot}=0$ and $\lim [\tilde{\vartheta}_i(t)]^{\cdot}=0$

Now, suppose that neither 1) $\tilde{b}^*_i(t)\to b^*_i$ and $\tilde{\vartheta}_i(t)\to\theta_i$ 2) $[[\tilde{b}^*_i(t)-b^*_i]^T \ [\tilde{\vartheta}_i(t)-\theta_i]^T]^T$ tends to be orthogonal to $\Omega_i(t)$. Suppose also that $\Omega_i(t)$ is not bounded. Then, it is clear that from eq.(8.20) $|\varepsilon_i(t)|$ will diverge with the same rate as $||\Omega_i(t)||$, which contradicts to the conclusion obtained in the previous paragraph. That is, even if $\Omega_i(t)$ is not bounded, either 1) $\tilde{b}^*_i(t)\to b^*_i$ and $\tilde{\vartheta}_i(t)\to\theta_i$ or 2) $[[\tilde{b}^*_i(t)-b^*_i]^T \ [\tilde{\vartheta}_i(t)-\theta_i]^T]^T$ tends to be orthogonal to $\Omega_i(t)$ must hold. Therefore, $\varepsilon_i(t)\to 0$ and hence both $[\tilde{b}^*_i(t)]^{\cdot}\to 0$ and $[\tilde{\vartheta}_i(t)]^{\cdot}\to 0$ are concluded regardless of the boundedness of $\Omega_i(t)$.

From the fact that both $[\tilde{b}^*_i(t)]^{\cdot}\to 0$ and $[\tilde{\vartheta}_i(t)]^{\cdot}\to 0$ hold and from eq.(8.17), we obtain

$$\tilde{e}_i(t) \to \frac{r^*_i(p)}{p^*_i(p)}\{[\tilde{b}^*_i(t)]^T u(t) + [\tilde{\vartheta}_i(t)]^T \omega_i(t)\} - y_{m,i}(t), \qquad (8.23)$$

where $y_{m,i}(t)$ is the i-th component of $y_m(t)$. Now, $[\tilde{b}^*_i(t)]^T u(t) + [\tilde{\vartheta}_i(t)]^T \omega_i(t)$ is equal to $g_{di}\bar{v}_i(t)$ from eq.(8.11), and hence eq.(8.23) implies that $\tilde{e}_i(t)\to 0$.

Since $\varepsilon(t)\to 0$ has been proven, $\tilde{e}(t)\to 0$ implies $e(t)\to 0$.

It remains to show that $\Omega(t)$ is bounded. again from the fact that both $[\tilde{b}^*_i(t)]^{\cdot}\to 0$ and $[\tilde{\theta}_i(t)]^{\cdot}\to 0$ hold and the fact that $\varepsilon(t)\to 0$, eq.(8.18) yields

$$\frac{r^*_i(p)}{p^*_i(p)}\{[\tilde{b}^*_i(t)]^T u(t)+\tilde{\theta}_i^T(t)\omega_i(t)-b^*_i{}^T u(t)-\theta^T\omega_i(t)\} \to 0. \tag{8.24}$$

Since $r^*_i(p)[p^*_i(p)]^{-1}$ is stable, eq.(8.24) implies

$$[\tilde{b}^*_i(t)]^T u(t)+\tilde{\theta}_i^T(t)\omega_i(t) \to b^*_i{}^T u(t) +\theta_i^T\omega_i(t), \tag{8.25}$$

which means that the plant input tends to the plant input that would be generated in the case of decoupling control with exact model matching which is of course bounded. Therefore, all signals within the systems are bounded.

REFERENCES

[1] R.V.Monopoli and C.C.Hsing; Parameter Adaptive control of multivariable systems, INT.J.Control,22,3,313/327, 1975.

[2] G.C.Goodwin, D.J.Ramadge, and P.E.Caines; Discrete-time multivariable adaptive control, IEEE Trans. Vol. AC-25,449/456,1980

[3] H.Elliott and W.A.Wolovich; A parameter adaptive control structure for linear multivariable systems,IEEE Trans. Vol.AC-27.340/352,1982.

[4] G.C.Goodwin and R.S.Long;Generalization of results on multivariable adaptive control, IEEE Trans. Vol.AC-29,761/764,1984.

Chapter 9 Discrete time system

9.1 Introduction.

9.1.1 Description of discrete time systems. We consider a scalar system for simplicity. The discrete time systems are described by difference equation, pulse transfer function and state vector difference equation, corresponding to differential equation, transfer function and state vector differential equation for continuous time systems respectively. That is,

$$y(k) = -a_1 y(k-1)-\cdots-a_n y(k-n)+b_{n-m}u(k-n+m) +b_n u(k-n), \tag{9.1}$$

$$t(z) = \frac{y(z)}{u(z)} = \frac{g(z^m+r_{m-1}z^{m-1}+\cdots+r_0)}{z^n+p_{n-1}z^{n-1}+\cdots+p_0}, \tag{9.2}$$

and

$$x(k) = Ax(k-1) + bu(k-1) \tag{9.3a}$$
$$y(k) = c^T x(k). \tag{9.3b}$$

9.1.2 Characteristic equation. If $u(k)=0$ in eq.(9.1), the system becomes free system; i.e., the system assumes a free motion starting from a given initial state. If we assume the form for the solution as $y(k)=\lambda^k$, the following equation must hold.

$$\lambda^k+a_1\lambda^{k-1}+\cdots+a_n=0 \tag{9.4}$$

This eqation is called a characteristic equation which has n roots. Assume that the characteristic roots are distinct, then the general solution is obtained as

$$y(k) = \sum_{i=1}^{n} c_i\lambda_i^k. \tag{9.5}$$

The constants c_i, $i=1,2,\cdots,n$ are determined when n initial conditions; e.g. $y(0),y(1),\cdots,y(n-1)$, are specified. When $\lim_{k\to\infty} y(k)=0$. the system is said asymptotically stable. The necessary and sufficient condition for asymptotic stability is that the magnitudes of all characteristic roots are less than 1. That is, as for stability, the left half plane for the continuous system correspods to the inside of unit circle.

9.1.4 z-transform. As is known, the Laplace transform of a time function $y(t)$ is defined by

$$y(s) = \int_0^\infty y(t)e^{-st}\, dt. \tag{9.6}$$

Similarly, the z-transform of a time series $\{y(k)\}$ is defined by

$$y(z) = \sum_{k=0}^{\infty} y(k)\, z^{-k}. \tag{9.7}$$

Table 9.1 shows Laplace transform of typical time functions and z-transform of time series obtained by sampling them with interval T. We denote the z-transform of a time series $\{\cdot\}$ by $Z\{\cdot\}$. Let $\{y(k)\}$ be a time series such that $y(k)=0$ for $k<0$ and denote $Z\{y(k)\}$ by $y(z)$, then we have $Z\{y(k-i)\}= \sum_{k=0}^{\infty} y(k-i)z^{-k}=z^{-i}y(z)$. The wave form of $\{y(k-i)\}$ shifts i steps to the right of $\{y(k)\}$;i.e., is retarded by i steps. This effect amouts the multiplication of z^{-i} in frequency domain. We thus can regard z^{-1} as a delay operator which retards a wave form by one step.

Assuming that both $y(k)$ and $u(k)$ are zero for $k<0$, the term by term z-transform of eq.(9.1) yields

Table 9.1 z transformation

wave form	time function $y(t)$	Laplace transfor $y(s)$	time series $y(k)$	z transform $y(z)$
impulse	$\delta(t)$	1	$\delta(k)$	1
step	$u(t)$	$1/s$	1	$1/(1-z^{-1})$
ramp	t	$1/s^2$	kT	$Tz^{-1}/(1-z^{-1})^2$
exponential	e^{-at}	$1/(s+a)$	e^{-kaT}	$1/(1-e^{-aT}z^{-1})$
sinusoidal	$\cos \omega t$	$s/(s^2+\omega^2)$	$\cos k\omega T$	$\dfrac{1-\cos \omega T\cdot z^{-1}}{1-2\cos \omega T\, z^{-1}+z^{-2}}$

$$y(z)=-a_1z^{-1}y(z)-\cdots-a_nz^{-n}y(z)+b_mz^{-(n-m)}u(z)$$
$$+\cdots+b_nz^{-n}u(z),$$

from which we obtain

$$\frac{y(z)}{u(z)} = \frac{b_mz^{-(n-m)}+\cdots+b_nz^{-n}}{1+a_1z^{-1}+\cdots+a_nz^{-n}} \tag{9.8a}$$

$$= \frac{b_mz^m+\cdots+b_n}{z^n+a_1z^{n-1}+\cdots+a_n}. \tag{9.8b}$$

Also, if $x(-1)=0$ and $u(-1)=0$, the z-transform of eq.(9.3) becomes

$$x(z)=Az^{-1}x(z)+bz^{-1}u(z)$$
$$y(z)=c^Tx(z),$$

from which we obtain

$$y(z)=c^T(I-Az^{-1})^{-1}z^{-1}bu(z)$$
$$=c^T(zI-A)^{-1}bu(z). \tag{9.9}$$

Therefore, the pulse transfer function is also described as $c^T(zI-A)^{-1}b$.

When $m=n-1$, we have

$$y(k)=-a_1y(k-1)-\cdots-a_ny(k-n)+b_1u(k-1)+\cdots+b_nu(k-n), \tag{9.10}$$

We say that the system (9.10) has no dead time, while the system (9.1) has dead time of $n-m-1$. Tho essence of dead time in discrete time systems, however, is quite differnt from that in continuous time systems. The dead time in discrete time systems are nothing but the relative degree of more than one. On the other hand, the dead time in continuous time systems alters a lumped parameter system to a distributed parameter one.

$\delta(k)$ decribed by Kronecker delta represents a wave form with a value 1 at $k=0$ and 0 elsewhere, and is called a unit impulse. The system response due to the input of

$\delta(k)$ with 0 initial conditions; i.e., $y(-1)=0$, $y(-2)=0$, $\cdots$ $y(-n)=0$, is called impulse response. The impulse response is determined as

$$y(0)=0;y(1)=b_1;y(2)=b_2-a_1y(1);\cdots;y(n)=b_n-a_1y(n-1)-$$
$$\cdots-a_{n-1}y(1);y(n+1)=-a_1y(n)-\cdots-a_{n-1}y(2)-a_ny(1);\cdots \tag{9.11}$$

Denote the solution by $\{t(k)\}$. The z-transform $t(z)$ is determined as

$$\begin{aligned}t(z)&=t(0)+t(1)z^{-1}+\cdots\\&=b_1z^{-1}+\cdots+b_nz^{-n}\\&\quad-a_1\{t(1)z^{-2}+t(2)z^{-3}+\cdots\}\\&\quad-a_2\{t(1)z^{-3}+t(2)z^{-4}+\cdots\}\\&\quad\cdots\cdots\cdots\cdots\cdots\cdots\\&=b_1z^{-1}+\cdots+b_nz^{-n}-a_1z^{-1}t(z)-\cdots-a_nz^{-n}t(z),\end{aligned}$$

from which

$$t(z)=\frac{b_1z^{-1}+\cdots+b_nz^{-n}}{1+a_1z^{-1}+\cdots+a_nz^{-n}}. \tag{9.12}$$

That is, the z-transform of impulse response is pulse transfer function, quite the same as continuous time systems.

Impulse response is also obtained from state space representation as shown below. Set $x(-1)=0$ and $u(k)=\delta(k)$, then we obtain $x(0)=0$ and $x(k)=A^{k-1}b$, $k=1,2,\cdots$. Therefore, the impulse response becomes

$$y(0)=0,\ y(k)=c^TA^{k-1}b,\ k=1,2,\cdots. \tag{9.13}$$

The z-transform of this is

$$\begin{aligned}y(z)&=c^Tbz^{-1}+c^TAbz^{-2}+\cdots+c^TA^kbz^{-k}+\cdots\\&=c^T(zI-A)^{-1}b,\end{aligned} \tag{9.14}$$

which is identical with the pulse transfer function.

The z-inverse transform of pulse transfer function t(z) yields the impulse response of the system. The z-inverse transform is obtained as the sequence of coefficients of the power series expansion of t(z) in z^{-1}, which is merely obtained by executing division algorithm after ordering both of numerator and denominator of t(z) in descent order.

9.1.5 Pulse transfer function of the plant. Assume the plant transfer function is given by t(s) because the plant itsel operates in continuous time. The output sequence {y(k)} is assumed to be obtained by sampling y(t) in a constant time interval T(sec). In general, some constant computing time T_c(sec) is needed to calculate the control signal u(k), where $0 \leq T_c \leq T$ is assumed. The calculated value of u(k) is held and is applied to the plant during T sec period from the instant T_c sec past the sampling instant. Such an operation is repeated in each T sec. (See Fig. 9.1).

Denote the step response in continuous time of the plant by $y_s(t)$, which can be calculated if the transfer function or state space representation of the plant is given. Since the stepwise waveform can be represented as the combination of step inputs, y(k), k=1,··· can be calculated as follows;

$$y(1)=y_s(T-T_c)u(0)$$

$$y(2)=y_s(2T-T_c)u(0)-y_s(T-T_c)u(0)+y_s(T-T_c)u(1)$$

$$y(3)=y_s(3T-T_c)u(0)-y_s(2T-T_c)u(0)$$
$$+y_s(2T-T_c)u(1)-y_s(T-T_c)u(1)+y_s(T-T_c)u(2)$$

$$\cdots\cdots\cdots\cdots\cdots\cdots\cdots\cdots\cdots\cdots . \quad (9.15)$$

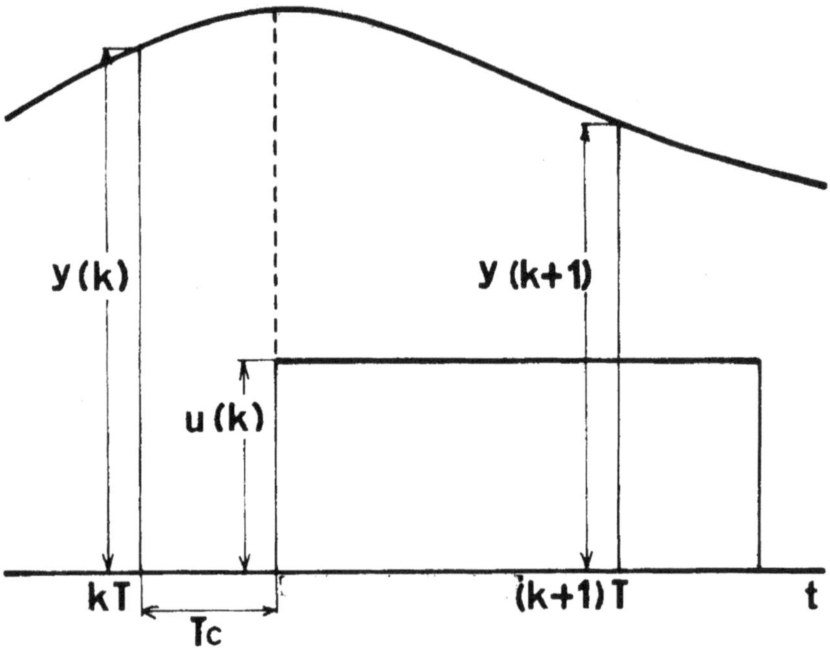

Fig. 9.1 Wave forms of y(k) and u(k)
in the presence of computing time

Therefore,

$$y(z) = u(z)(1-z^{-1})\ Z\{y_s(kT-T_c)\}, \tag{9.16}$$

i.e.,

$$\text{pulse transfer function} = (1-z^{-1})\ Z\{y_s(kT-T_c)\}. \tag{9.17}$$

When the continuous time plant transfer function has dead time L, the step response becomes $y_s(t-L)$, where $y_s(t)$ is the step response of the plant without dead time. The pulse transfer function can be determined by eq.(9.17) again, but two cases occur according to relativity of T_c and L. As an example, consider a plant described by $t(s)=K/(s+\lambda)$. The pulse transfer function is

$$t(z)=K/\lambda\cdot(1-e^{-\lambda T})z^{-1}/(1-e^{\lambda T}z^{-1}).$$

Suppose then the plant has dead time L. Let define an integer n_L by $L=n_LT+T_L$, where $0\leq T_L\leq T$. The pulse transfer function when $T_c<T-T_L$ is

$$t(z)=K/\lambda\cdot\{(1-e^{\lambda(T_L+T_c)}\ e^{-\lambda T})z^{-(n_L+1)}-(1-e^{\lambda(T_L+T_c)})\\ \times e^{-\lambda T}z^{-(n_L+2)}\}/(1-e^{-\lambda T}z^{-1}),$$

while the pulse transfer function when $T-T_L\leq T_c<T$ is

$$t(z)=K/\lambda\cdot\{(1-e^{\lambda(T_L+T_c-T)}e^{-\lambda T})z^{-(n_L+2)}-(1-e^{\lambda(T_L+T_c-T)})\\ \times e^{-\lambda T}z^{-(n_L+3)}\}/(1-e^{-\lambda T}z^{-1}).$$

The example demonstrates the fact that the relative degree of $t(z)$ is 1 when the dead time is zero, but when dead time is not zero, it may be n_L when T_c is short or n_L+1 when T_c is long.

9.2 Pole assignment. Let the plant pulse transfer function be

$$t(z)=gr(z)p^{-1}(z), \tag{9.18}$$

where r(z) and p(z) are m and n degree monic polynomials respectively with $0 \leq m \leq n-1$. Also, r(z) and p(z) are assumed to be relatively prime. Arbitrary pole assignment is to construct a feedback system so that the closed loop characteristic polynomial is identical to an arbitrarily given n degree monic polynomial $p_f(z)$. Fortunately quite the same theory as continuous time systems can be applied. Let $f(z)=p(z)-p_f(z)$ and introduce any n-1 degree monic stable polynomial q(z). The n-2 and n-1 degree polynomials k(z) and H(z) respectively can be determined uniquely from the polynomial equation

$$k(z)p(z)+h(z)gr(z)=q(z)f(z). \tag{9.19}$$

Then, take a control law as

$$u(z)=\frac{k(z)}{q(z)}u(z) + \frac{h(z)}{q(z)}y(z) + v(z), \tag{9.20}$$

where v(k) is an external reference input. Then the specified pole assignment is achieved.

Caution must be paid to the fact that the control function in the discrete time system is not parallel as is in the continuous time system but is sequential. Let us represent q(z), k(z) and h(z) as shown in the following.

$$q(z)=z^{n-1}+q_{n-2}z^{n-2}+\cdots+q_0 \tag{9.21}$$

$$k(z)= \quad k_{n-2}z^{n-2}+\cdots+k_0 \tag{9.22}$$

$$h(z)=h_{n-1}z^{n-1}+h_{n-2}z^{n-2}+\cdots+h_0 \tag{9.23}$$

Then the control law (9.20) can be written in time domain as follows.

$$\begin{aligned}u(k)=&-q_{n-2}u(k-1)-\cdots-q_0u(k-n+1)\\&+k_{n-2}u(k-1)+\cdots+k_0u(k-n+1)\\&+h_{n-1}y(k)+h_{n-2}y(k-1)+\cdots+h_0y(k-n+1)\\&+v(k)+q_{n-2}v(k-1)+\cdots+q_0v(k-n+1)\end{aligned} \tag{9.24}$$

The existence of $y(k)$ and $v(k)$ in the right hand side of eq.(9.24) means that $y(k)$ and $v(k)$ are detected at the instant k, $u(k)$ is computed by eq.(9.24) spending $T_c$sec, and then the resulting $u(k)$ is applied to the plant. All other signals are available from the memory of the computer.

9.3 Exact model matching. Let the pulse transfer function of the reference model be given by

$$t_d(z)=g_d r(z)p_d^{-1}(z), \tag{9.25}$$

where $r_d(z)$ and $p_d(z)$ are m_d and n_d degree monic polynomials respectively. The following assumptions are needed.

A.1 $r(z)$ is stable.

A.2 $n_d-m_d \geq n-m$

Again, the same theory as in continuous time systems can be applied. By introducing any m and n degree monic stable polynomials $r^*(z)$ and $p^*(z)$ respectively, we define an input dynamics by

$$t_{IN}(z)=\frac{r_d(z)p^*(z)}{r^*(z)p_d(z)} . \tag{9.26}$$

Also define $\bar{v}(z)=t_{IN}(z)v(z)$. Let $\gamma(z)$ be any n-m-1 degree monic stable polynomial. The n-2 and n-1 degree polynomials $k_b(z)$ and $h(z)$ respectively can be determined uniquely from the polynomial equation

$$k_b(z)p(z)+h(z)r(z)=\gamma(z)r^*(z)p(z)-\gamma(z)r(z)p^*(z). \tag{9.27}$$

The control law is specified as

$$u(z)=\frac{1}{g}\{\frac{k(z)}{\gamma(z)r^*(z)}u(z)+\frac{h(z)}{\gamma(z)r^*(z)}\}+\frac{g_d}{g}\bar{v}(z), \tag{9.28}$$

where $k(z)=gk_b(z)$. Then the specified exact model matching is achieved. Denote the polynomial $\gamma(z)r^*(z)$ by

$$\tau(z)r^*(z)=z^{n-1}+\tau_{n-2}z^{n-2}+\cdots+\tau_0. \tag{9.29}$$

The signal $\bar{v}(k)$ is generated by the following equation.

$$\begin{aligned}\bar{v}(k)=&-a_1\bar{v}(k-1)-\cdots-a_{n_d+m}\bar{v}(k-n_d-m)\\ &+v(k-n_d-m+n+m_d)+b_1v(k-n_d-m+n+m_d-1)\\ &+\cdots+b_{n+m_d}v(k-n_d-m),\end{aligned} \tag{9.30}$$

where a_i and b_i are the coefficients of the polynomials $r^*(z)p_d)z)$ and $r_d(z)p^*(z)$ respectively. On the other hand, the control law is given by

$$\begin{aligned}u(k)=&-\tau_{n-2}u(k-1)-\cdots-\tau_0u(k-n+1)\\ &+(1/g)\cdot\{k_{n-2}u(k-1)+\cdots+k_0u(k-n+1)\\ &+h_{n-1}y(k)+h_{n-2}y(k-1)+\cdots+h_0y(k-n+1)\}\\ &+(g_d/g)\cdot\{\bar{v}(k)+\tau_{n-2}\bar{v}(k-1)+\cdots+\tau_0\bar{v}(k-n+1)\}.\end{aligned} \tag{9.31}$$

9.4 Design of finite time settling system. When the error extincts within finite time (stages) in the step or ramp responses, the system is called finite time settling system. Since the phenomenon does not observed in linear continuous time control system, it has been considered as a peculiar problem. Various explanations about the design of finite time settling systems appear in the literature [1], but it is to be noticed that the design problem is nothing but the appropriate choice of the reference model in the exact model matching problem.

In the finite time settling problem, the exterior reference input is step or ramp signal, and the initial conditions are assumed to be zero. Since $e(z)=\sum_{k=0}^{\infty} e(k)z^{-k}$, finite time settling means that $e(z)$ should be represented by q degree polynomial in z^{-1} for some finite number q. Therefore, it is clear that the exact model matching for the reference model of the form

$$t_d(z) = 1 - \frac{\sum_{k=0}^{\infty} g_k z^{-k}}{v(z)} \qquad (9.33)$$

achieves finite settling time control, where $v(z)=1/(1-z^{-1})$ for step input and $v(z)=z^{-1}/(1-z^{-1})^2$. q and g_k, $k=0,1,\cdots,q$ are the design parameters. As was pointed earlier, the relative degree of the reference model must not be less than that of the plant, which imposes a constraint on the value of g_d. Let the relative degree of the plant be d.

<u>Step response finite time settling</u>. Since $v(z)=1/(1-z^{-1})$, eq.(6.33) becomes

$$t(z)=\frac{(1-g_0)z^{q+1}+(g_0-g_1)z^q+\cdots+(g_{q-1}-g_q)z+g_q}{z^{q+1}}. \qquad (9.34)$$

Since the degree of the numerator must be at most q+1--d, we obtain

$$1-g_0=0,\ g_0-g_1=0,\ \cdots,g_{d-2}-g_{d-1}=0, \qquad (9.35)$$

from which $g_0=g_1=\cdots=g_{d-1}=1$ are obtained. Therefore, the reference model must be

$$t(z)=\frac{(1-g_d)z^{q+1-d}+\cdots+(g_{q-1}-g_q)z+g_q}{z^{q+1}}. \qquad (9.36)$$

We from eq.(9.36) that q should not be less then d. The resulting e(k) and y(k) are obtained as follows.

$$e(k)=\{1,\cdots,1,g_d,\cdots,g_q,0,0,\cdots\} \qquad (9.37)$$

$$y(k)=\{0,\cdots,0,1-g_d,\cdots,1-g_q,1,1,\cdots\} \qquad (9.38)$$

<u>Ramp response finite settling</u>. Since $v(z)=z^{-1}/(1-z^{-1})^2$, we obtain

$$t_d(z)=\frac{\left(\begin{array}{l}-g_0z^{q+2}+(1+2g_0-g_1)z^{q+1}+\cdots\\+(-g_{q-2}+2g_{q-1}-g_q)z^2\\+(-g_{q-1}+2g_q)z-g_q\end{array}\right)}{z^{q+1}} \qquad (9.39)$$

We now obtain

$$-g_0=0,\ 1+2g_0-g_1=0, -g_0+2g_1-g_2=0,\cdots,$$
$$-g_{d-2}+2g_{d-1}-g_d=0, \qquad (9.40)$$

from which we obtain $g_k=k$, $k=0,1,\cdots,d$. Hence,

$$t_d(z)=\frac{\left(\begin{array}{l}(d+1-g_{d+1})z^{q+1-d}+(-d+2g_{d+1}-g_{d+2})z^{q-d}\\+(-g_{d+1}+2g_{d+2}-g_{d+3})z^{q-d-1}+\cdots\\+(-g_{q-1}+2g_q)z-g_q\end{array}\right)}{z^{q+1}}. \qquad (9.41)$$

We know that q should not be less than d+1. The resulting e(k) and y(k) are as follows.

$$e(k)=\{0,1,\cdots,d,g_{d+1},\cdots,g_q,0,0,\cdots\} \qquad (9.42)$$
$$y(k)=\{0,0,\cdots,0,d+1-g_{d+1},\cdots,q-g_q,q+1,q+2,\cdots\} \qquad (9.43)$$

9.5 Adaptive control.[2]-[5] The adaptive control of discrete time systems has been cosidered to be peculiar, and rather complicated methods were reported formerly [2]. As was illustrated above, the control theory for discrete time systems is almost the same as that for continuous time systems. The reason why the discrete time systems look like peculiar comes from the improper timing. Many operations such as detection of output y(k), updating of parameters, computation of input u(k) using the updated parameters, etc., must be carried out in right sequence. Going through in right sequence yields the simple ad-aptive control as continuous time systems.

9.5.1 Adaptive identification. The motion of the plant is assumed to be represented by eq.(9.1), but it is assumed that m=n-1. Put

$$\theta=[-a_1,\cdots,-a_n,b_1,\cdots,b_n]^T \qquad (9.44)$$

$$\omega(k-1)=[y(k-1),\cdots,y(k-n),u(k-1),\cdots,u(k-n)]^T \qquad (9.45)$$

Then, eq.(9.1) can be rewritten as

$$y(k)=\theta^T\omega(k-1), \qquad (9.46)$$

where θ is unknown parameter vector, while $\omega(k-1)$ is known and hence available signal. The reason to denote $\omega(\cdot)$ by $\omega(k-1)$ is to make clear that all signals in $\omega(\cdot)$ are ready before the instant k. Define an adaptive identifier by

$$\tilde{y}(k)=\tilde{\theta}^T(k-1)\omega(k-1). \qquad (9.47)$$

The reason to denote $\tilde{\theta}(\cdot)$ by $\tilde{\theta}(k-1)$ is that the parameters are not yet revised at the instant k. y(k) is detected at the instant k. These parameters are not revised to $\tilde{\theta}(k)$ until the compution process based on adaptive algorithm mentioned below is finished. The identification error is thus

$$e(k)=\tilde{y}(k)-y(k)=[\tilde{\theta}(k-1)-\theta]^T\omega(k-1). \qquad (9.48)$$

Adaptive law is the prescription of the rule to revise $\tilde{\theta}(k-1)$ to $\tilde{\theta}(k)$ so as to make e(k) tend to zero, although the ultimate object of identification is to make $\tilde{\theta}(k)$ tend to θ. Since the available signals are e(k) and $\omega(k-1)$, the adaptive law should be prescribed in terms of e(k) and $\omega(k-1)$.

There are many variations for adaptive law, but we consider in the first an adaptive law based on the lest square principle. Although the plant dynamics is represented by eq.(9.46), we introduce an observation error ε (k); i.e., we assume

$$y(i) = \theta^T\omega(i-1) + \varepsilon(i), \quad i=1,2,\cdots,k. \tag{9.49}$$

The least square principle means that $\tilde{\theta}(k)$, the estimate of θ, should be determined so that $\sum_{i=1}^{k} \varepsilon^2(i)$ be minimum. By differentiating $\sum_{i=1}^{k}[y(i)-\theta^T\omega(k-1)]^2$ by θ and setting the derivative equal to zero, $\tilde{\theta}(k)$ is obtained as

$$\tilde{\theta}(k) = [\sum_{i=1}^{k} \omega(i-1)\omega^T(i-1)]^{-1}[\sum_{i=1}^{k} \omega(i-1)y(i)], \tag{9.50}$$

where $\sum_{i=1}^{k} \omega(i-1)\omega^T(i-1)$ is assumed to be positive definite.

From eq.(9.50), we obtain

$$[\sum_{i=1}^{k} \omega(i-1)\omega^T(i-1)]\tilde{\theta}(k)= \sum_{i=1}^{k} \omega(i-1)y(i). \tag{9.51}$$

On the other hand, if we apply the relation (9.50) to the instant k-1, we obtain

$$[\sum_{i=1}^{k} \omega(i-1)\omega^T(i-1)-\omega(k)\omega^T(k)]\tilde{\theta}(k-1)= \sum_{i=1}^{k-1} \omega(i-1)y(i). \tag{9.52}$$

The above two equations yield

$$[\sum_{i=1}^{k} \omega(i-1)\omega^T(i-1)](\tilde{\theta}(k)-\tilde{\theta}(k-1))=\omega(k-1)y(k) -\omega(k-1)\omega^T(k-1)\tilde{\theta}(k-1). \tag{9.53}$$

By using the relation (9.47), the right hand side of the above equation can be written as $-\omega(k-1)e(k)$. Then, we obtain the following as an adaptive law.

$$\tilde{\theta}(k)=\tilde{\theta}(k-1)-[\sum_{i=1}^{k} \omega(i-1)\omega^T(i-1)]^{-1}\omega(k-1)e(k). \tag{9.54}$$

In order to avoid operations of both inverse and summation, we introduce a matrix $\Gamma(k)$ by

$$\Gamma(k) = [\sum_{i=1}^{k} \omega(i-1)\omega^T(i-1)]^{-1}. \tag{9.55}$$

Then, the adaptive law can be written as

$$\tilde{\theta}(k) = \tilde{\theta}(k-1) - \Gamma(k)\omega(k-1)e(k). \tag{9.56}$$

On the other hand, we obtain from eq.(9.55)

$$\Gamma^{-1}(k) = \Gamma^{-1}(k-1) + \omega(k-1)\omega^T(k-1). \tag{9.57}$$

By applying matrix inversion lemma to eq.(9.57), we obtain

$$\Gamma(k) = \Gamma(k-1) - \frac{\Gamma(k-1)\omega(k-1)\omega^T(k-1)\Gamma(k-1)}{1 + \omega^T(k-1)\Gamma(k-1)\omega(k-1)}. \tag{9.58}$$

Since $\Gamma^{-1}(0)=0$, $\Gamma(0)$, the initial conditions for $\Gamma(k)$, should be infinity ideally. In reality, $\Gamma(0)$ is set to αI with α being a sufficiently large number. Thus , the adaptive law based on the least square principle is given by

$$\left.\begin{aligned} \tilde{\theta}(k) &= \tilde{\theta}(k-1) - \Gamma(k)\omega(k-1)e(k-1) \\ \Gamma(k) &= \Gamma(k-1) - \frac{\Gamma(k-1)\omega(k-1)\omega^T(k-1)\Gamma(k-1)}{1 + \omega^T(k-1)\Gamma(k-1)\omega(k-1)} \\ \Gamma(0) &= \alpha I,\ \alpha >> 1 \end{aligned}\right\}. \tag{9.59}$$

Theorem 9.1 (adaptive law-1) Assume that $\omega(k)$ is bounded. Then by an adaptive law (9.59), $e(k)\to 0$, $\phi(k)\omega(k)\to 0$ and $\tilde{\theta}(k)-\tilde{\theta}(k-1)\to 0$, where $\phi(k)=\tilde{\theta}(k)-\theta$.

(proof) As a candidate of Liapunov function, we introduce a function

$$V(k) = \phi^T(k)\Gamma^{-1}(k)\phi(k). \tag{9.60}$$

Since $\Gamma^{-1}(0)$ is positive definite and $\omega(k-1)\omega^T(k-1)$ is nonnegative, $\Gamma^{-1}(k)$ is positive definite. Therefore, $V(k)$ cannot be negative. Also, $V(k)$ becomes infinity if and only if $||\phi(k)||$ is infinite.

On the other hand, we obtain

$$\begin{aligned} \Delta V(k) &= \phi^T(k)\Gamma^{-1}(k)\phi(k) - \phi^T(k-1)\Gamma^{-1}(k-1)\phi(k-1) \\ &= [\phi(k-1)-\Gamma(k)\omega(k-1)e(k)]^T\ \Gamma^{-1}(k) \end{aligned}$$

$$\times[\phi(k-1)-\Gamma(k)\omega(k-1)e(k)] -\phi^T(k-1)\Gamma^{-1}(k-1)\phi(k-1)$$
$$= \phi^T(k-1)[\Gamma^{-1}(k)-\Gamma^{-1}(k-1)]\phi(k-1) -2\phi^T(k-1)\omega(k-1)e(k)$$
$$+ e^2(k)\omega^T(k-1)\Gamma(k)\omega(k-1)$$
$$= -e^2(k)\{1 - \omega^T(k-1)\Gamma(k)\omega(k-1)\}.$$

However, from eq.(9.58)

$$\omega^T(k-1)\Gamma(k)\omega(k-1) = \frac{\omega^T(k-1)\Gamma(k-1)\omega(k-1)}{1+\omega^T(k-1)\Gamma(k-1)\omega(k-1)}.$$

Thus, $\Delta V(k)$ becomes

$$\Delta V(k) = -e^2(k)\frac{1}{1+\omega^T(k-1)\Gamma(k-1)\omega(k-1)} \leq 0. \tag{9.61}$$

Therefore, $\phi(k)$ must be bounded. Since $V(k)>0$ and $\Delta V(k) \leq 0$, $V(k)$ decreases monotonically until it converges to some nonnegative constant, hence $\Delta V(k)$ converges to zero. Suppose that $\omega(k)$ is uniformly bounded. Then, $e(k)$ converges to zero. Suppose then that $\omega(k)$ diverges to infinity. Since $\Delta(k)\to 0$, $|e(k)|$, if it would diverge, grows with lower rate than $||\omega(k)||$, which implies that either 1) $\phi(k)$ tends to zero or 2) $\phi(k)$ tends to be orthogonal to $\omega(k)$. For both cases, $e(k)$ must tends to zero. That is, $e(k)$ always tends to zero regardless of the boundedness of $\omega(k)$. Then, $\phi^T(k)\omega(k)$ converges to zero. Now, $\Gamma(k)$ is bounded, and $\omega(k)$ is bounded by assumption. Therefore, from eq.(9.59), we obtain $\tilde{\theta}(k)-\tilde{\theta}(k-1)\to 0$. Q.E.D.

The following adaptive law is derived from Liapunov stability theory.

$$\left.\begin{array}{l} \tilde{\theta}(k) = \tilde{\theta}(k-1) - \dfrac{\alpha\Gamma e(k)\omega(k-1)}{c + \omega^T(k-1)\Gamma\omega(k-1)} \\ \Gamma=\Gamma^T>0,\ c>0,\ 2>\alpha>2 \end{array}\right\}. \tag{9.62}$$

We do not assume the boundedness of $\omega(k)$ in this law.

Theorem 9.2 (adaptive law-2) By the adaptive law (9.62), $e(k)\to 0$, $\phi^T(k)\omega(k)\to 0$ and $\tilde{\theta}(k)-\tilde{\theta}(k-1)\to 0$.

(proof) As a candidate of Liapunov function, we introduce a function

$$V(k) = \phi^T(k)\Gamma^{-1}\phi(k). \tag{9.63}$$

$V(k)$ is nonnegative and becomes infinity if and only if $\|\phi(k)\|$ is infinite. Now,

$$\begin{aligned}\Delta V(k) &= \frac{e^2(k)}{c+\omega^T(k-1)\Gamma\omega(k-1)}\left\{-2 + \alpha\,\frac{\omega^T(k-1)\Gamma\omega(k-1)}{c+\omega^T(k-1)\Gamma\omega(k-1)}\right\}\\ &\le \frac{e^2(k)}{c+\omega^T(k-1)\Gamma\omega(k-1)}(-2+\alpha) \le 0,\end{aligned} \tag{9.64}$$

since $2>\alpha>0$. Therefore, $\phi(k)$ must be bounded. Quite the same reason as the proof for the previous theorem, $e(k)$ as well as $\phi^T(k)\omega(k)$ converges to zero regardless of the boundedness of $\omega(k)$. Furthermore, eq.(9.62) assures $\tilde{\theta}(k)-\tilde{\theta}(k-1)\to 0$ regardless of the boundedness of $\omega(k)$. Q.E.D.

9.5.2 Adaptive control. Assumptions A.1 and A.2 are needed as for exact model matching. Also, input dynamics $t_{IN}(z)$ is introduced. We know that the control law when the plant is known is given by eq.(9.28). Put

$$k(z)= k_{n-2}z^{n-2}+\cdots+k_1z+k_0 \tag{9.65}$$

$$h(z)=h_{n-1}z^{n-1}+h_{n-2}z^{z-2}+\cdots+h_1z+h_0, \tag{9.66}$$

and then define

$$\theta=[-k_{n-2},\cdots,-k_0,-h_{n-1},-h_{n-2},\cdots,-h_0]^T \tag{9.67}$$

$$\omega(k)=[\frac{z^{n-2}}{\tau(z)r^*(z)}u(k),\cdots,\frac{1}{\tau(z)r^*(z)}u(k),\frac{z^{n-1}}{\tau(z)r^*(z)}y(k),\cdots,\frac{1}{\tau(z)r^*(z)}y(k)]^T. \tag{9.68}$$

Then, the control law for exact model matching can be represented as

$$u(k)=-(1/g)\theta^T\omega(k)+(g_d/g)\bar{\upsilon}(k), \qquad (9.69)$$

where each component of $\omega(k)$ is determined as follows. Denote $\tau(z)r^*(z)$ as

$$\tau(z)r^*(z)=z^{n-1}+\tau_{n-2}z^{n-2}+\cdots+\tau_1 z+\tau_0. \qquad (9.70)$$

Then,

$$\left.\begin{array}{l}\omega_1(k)=-\tau_{n-2}\omega_1(k-1)-\cdots-\tau_0\omega_1(k-n+1)+u(k-1)\\ \omega_{i+1}(k)=\omega_i(k-1),\ i=1,2,\cdots,n-2\\ \omega_n(k)=-\tau_{n-2}\omega_n(k-1)-\cdots-\tau_0\omega_n(k-n+1)+y(k)\\ \omega_{j+1}(k)=\omega_j(k-1),\ j=n,n+1,\cdots,2n-2\end{array}\right\} \qquad (9.71)$$

Clearly, $\omega(k)$ is available immediately after $y(k)$ is detected, and does not contain $u(k)$.

Since g and θ are unknown in adaptive control, the control law for adaptive control is fixed to

$$u(k)=-(1/\tilde{g}(k))\tilde{\theta}^T(k)\omega(k)+(1/\tilde{g}(k))g_d\bar{\upsilon}(k), \qquad (9.72)$$

by using the parameter estimates to be defined later. Employing k for $\tilde{g}(\cdot)$ and $\tilde{\theta}(\cdot)$ means that the control $u(k)$ should be calculated after the completion of updating of $\tilde{g}(k-1)\to\tilde{g}(k)$ and $\tilde{\theta}(k-1)\to\tilde{\theta}(k)$.

Now, the plant dynamics should be represented in terms of $\{g,\ k(z),\ h(z)\}$. From eq.(9.27), we obtain

$$\frac{k_b(z)}{\tau(z)p^*(z)}u(z)+\frac{h(z)}{\tau(z)p^*(z)}\,\frac{1}{g}y(z)=\frac{r^*(z)}{p^*(z)}u(z)-\frac{1}{g}y(z), \qquad (9.73)$$

from which

$$y(k)=g\frac{r^*(z)}{p^*(z)}u(k)-\frac{k(z)}{\tau(z)r^*(z)}\,\frac{r^*(z)}{p^*(z)}u(k)$$
$$-\frac{h(z)}{\tau(z)r^*(z)}\,\frac{r^*(z)}{p^*(z)}y(k). \qquad (9.74)$$

Using further the notations for θ and $\omega(k)$, it is written as

$$y(k)=g\{\frac{r^*(z)}{p^*(z)}u(k)\} + \theta^T\{\frac{r^*(z)}{p^*(z)}\omega(k)\}. \tag{9.75}$$

Put

$$\left.\begin{aligned} \xi_u(k)&=\frac{r^*(z)}{p^*(z)}u(k) \\ \xi_\omega(k)&=\frac{r^*(z)}{p^*(z)}\omega(k) \end{aligned}\right\}. \tag{9.76}$$

Then eq.(9.73) can be written as

$$y(k)=g\xi_u(k) + \theta^T\xi_\omega(k), \tag{9.77}$$

where $\xi_u(k)$ and $\xi_\omega(k)$ are generated respectively by the folowing equations.

$$\left.\begin{aligned} \xi_u(k)=&-p^*_{n-1}\xi_u(k-1)-\cdots-p^*_0\xi_u(k-n) \\ &+u(k-n+m)+r^*_{m-1}u(k-n+m-1)+\cdots+r^*_0u(k-n) \\ \xi_\omega(k)=&-p^*_{n-1}\xi_\omega(k-1)-\cdots-p^*_0\xi_\omega(k-n) \\ &+\omega(k-n+m)+r^*_{m-1}\omega(k-n+m-1)+\cdots+r^*_0\omega(k-n) \end{aligned}\right\} \tag{9.78}$$

The adaptive control error is defined as $e(k)=y(k)-y_m(k)$; i.e.,

$$e(k)=g\xi_u(k) + \theta^T\xi_\omega(k) - y_m(k). \tag{9.79}$$

The eq.(9.79) represents a dynamics of adaptive control error system. The identifier for the error system is introduced as follows.

$$\tilde{e}(k)=\tilde{g}(k-1)\xi_u(k) + \tilde{\theta}^T(k-1)\xi_\omega(k) - y_m(k). \tag{9.80}$$

The identification error $\varepsilon(k)=\tilde{e}(k)-e(k)$ is obtained as

$$\varepsilon(k)=[\tilde{g}(k-1)-g \quad \tilde{\theta}^T(k-1)-\theta^T]\begin{bmatrix}\xi_u(k)\\ \xi_\omega(k)\end{bmatrix}. \tag{9.81}$$

Employing k-1 for $\tilde{g}(\cdot)$ and $\tilde{\theta}(\cdot)$ means that parameter values before updating are used in identification error. The adaptive law is determined as follows.

$$\begin{bmatrix}\tilde{g}(k)\\ \tilde{\theta}(k)\end{bmatrix}=\begin{bmatrix}\tilde{g}(k-1)\\ \tilde{\theta}(k-1)\end{bmatrix}-\frac{\alpha\Gamma\varepsilon(k)}{c+\begin{bmatrix}\xi_u(k)\\ \xi_\omega(k)\end{bmatrix}^T\begin{bmatrix}\xi_u(k)\\ \xi_\omega(k)\end{bmatrix}}\begin{bmatrix}\xi_u(k)\\ \xi_\omega(k)\end{bmatrix}, \qquad (9.82)$$

where $\Gamma=\Gamma^T>0$, $c>0$, $0<\alpha<2$.

For brevity, we put

$$\Phi(k)=[\tilde{g}(k)-g \quad \tilde{\theta}^T(k)-\theta^T]^T \qquad (9.83)$$

$$\Omega(k)=[\xi_u(k) \quad \xi_\omega{}^T(k)]^T . \qquad (9.84)$$

Then eqs.(9.81) and (9.82) are written respectively as follows.

$$\varepsilon(k)=\Phi^T(k)\Omega(k) \qquad (9.85)$$

$$\Phi(k)=\Phi(k-1)-\frac{\alpha\Gamma\varepsilon(k)\Omega(k)}{c+\Omega^T(k)\Gamma\Omega(k)} . \qquad (9.86)$$

As a candidate of Liapunov function, we take

$$V(k)=\frac{1}{\alpha}\Phi^T(k)\Gamma^{-1}\Phi(k).$$

Then, we obtain as the argument in 9.5.1

$$\Delta V(k)\leq \frac{\varepsilon^2(k)}{c+\Omega^T(k)\Gamma\Omega(k)}(-2+\alpha) \leq 0 .$$

It is seen that since $V(k)$ decreases monotonically, $||\Phi(k)||$ is bounded. Also $V(k)$ converges to some constant value, while $\Delta V(k)$ converges to zero. If $||\Omega(k)||$ is not bounded, the divergence speed of $|\varepsilon(k)|$, if it diverges, must be of small order of that of $||\Omega(k)||$. Therefore, either $\Phi(k)\to 0$ or $\Phi(k)$ orthogonalize to $\Omega(k)$ as $k\to\infty$, which implies $\varepsilon(k)\to 0$ as $k\to\infty$. Hence, from eq.(9.86) $\Delta\Phi(k)=\Phi(k)-\Phi(k-1)$ converges to zero. That is,

$$\left.\begin{aligned}&\lim_{k\to\infty}[\tilde{g}(k)-\tilde{g}(k-1)] = 0\\ &\lim_{k\to\infty}[\tilde{\theta}(k)-\tilde{\theta}(k-1)] = 0\end{aligned}\right\} \qquad (9.87)$$

A lemma is provided here.

Lemma 9.1 If eq.(9.85) holds, the following is true.

$$\lim_{k\to\infty} \tilde{g}(k-1)\{\frac{r^*(z)}{p^*(z)}u(k)\}=\lim_{k\to\infty} \frac{r^*(z)}{p^*(z)}\{\tilde{g}(k)u(k)\}$$

$$\lim_{k\to\infty} \tilde{\theta}^T(k-1)\{\frac{r^*(z)}{p^*(z)}\omega(k)\}=\lim_{k\to\infty} \frac{r^*(z)}{p^*(z)}\{\tilde{\theta}^T(k)\omega(k)\}.$$

(proof) Only the first half is proven. Since $\lim_{k\to\infty} [\tilde{g}(k)-\tilde{g}(k-1)]=0$, $\lim_{k\to\infty} [\tilde{g}(k)-\tilde{g}(k-i)]=0$, $i=1,2,\cdots$. Then, as $k\to\infty$

$$\tilde{g}(k-1)\xi_u(k)=-p^*_{n-1}\tilde{g}(k-2)\xi_u(k-1)-\cdots-p^*_0\tilde{g}(k-n-1)\xi_u(k-n)$$
$$+\tilde{g}(k-n+m)u(k-n+m)+\cdots+r^*_0\tilde{g}(k-n)u(k-n)$$

is obtained. The above equations is written as

$$\tilde{g}(k-1)\xi_u(k)=\frac{r^*(z)}{p^*(z)}\{\tilde{g}(k)u(k)\}.$$

That is the first equation is obtained.

Q.E.D.

Applying lemma 9.1 to eq.(9.80), we obtain

$$\lim_{k\to\infty} \tilde{e}(k)=\lim_{k\to\infty} [\tilde{g}(k-1)\xi_u(k)+\tilde{\theta}^T(k-1)\xi_\omega(k)-y_m(k)]$$
$$=\lim_{k\to\infty} \frac{r^*(z)}{p^*(z)}[\tilde{g}(k)u(k)+\tilde{\theta}^T(k)\omega(k)-y_m(k)],$$

which will be zero because of eq.(9.72). This together with the fact $\varepsilon(k)\to 0$ as $k\to\infty$ implies $e(k)\to 0$ as $k\to\infty$.

Furthermore, from $\varepsilon(k)\to 0$, i.e., $\Phi^T(k)\Omega(k)\to 0$,

$$\lim_{k\to\infty} (\tilde{g}(k)-g)\frac{r^*(z)}{p^*(z)}u(k)-(\tilde{\theta}(k)-\theta)^T \frac{r^*(z)}{p^*(z)}\omega(k))=0,$$

i.e.,

$$\lim_{k\to\infty} \frac{r^*(z)}{p^*(z)}[(\tilde{g}(k)-g)-(\tilde{\theta}(k)-\theta)^T\omega(k)]=0.$$

This implies $\lim_{k\to\infty} (\tilde{g}(k)-g)u(k)+(\tilde{\theta}(k)-\theta)^T\omega(k)=0$ because $r^*(z)/p^*(z)$ is a stable filter. The last equation means

$$\lim_{k\to\infty} u(k)=-\frac{1}{g}\ \theta^T\omega(k)+(g_d/g)\bar{v}(k).$$

This implies that the plant input u(k) in the adaptive

control converges to the plant input in the exact model matching which surely bounded, and the adaptive control system does work well.

It is important to establish a correct sequence of operations in each step. The sequence is shown below.

(1) detection of $y(k)$, $y_m(k)$, and $v(k)$.

(2) calculation of $e(k)$ and $\bar{v}(k)$.

(3) calculation of $\omega(k)$.

(4) calculation of $\xi_u(k)$ and $\xi_\omega(k)$.

(5) calculation of $\tilde{e}(k)$.

(6) calculation of $\varepsilon(k)$

(7) calculation of $\tilde{g}(k)$ and $\tilde{\theta}(k)$. (parameter updating)

(8) calculation of $u(k)$ and its application to the plant.

REFERENCES

[1] H. Kimura; Digital signal reduction and control, (in Japanese), Shokodo, 1982.

[2] S. Suzuki and S. Takashima; Design of a hyperstable discrete model reference adaptive control system, Trans. SICE, 13, 433/438, 1977.

[3] K.S. Narendra and Y.H. Lin; " Design of stable model reference adaptive controllers" in Applications of Adaptive Control ed. by Narendra and Monopoli, Academic Press, 1980.

[4] I.D. Landau and M. Tomizuka; Theory and practice of adaptive control systems, Ohm-sha, 1982.

[5] G.C.Goodwin, P.J.Ramadge and P.E.Caines; Discrete-time multivariable adaptive control, IEEE Trans.Vol.AC-25, 491/456, 1980.

Chapter 10 Time delay system

10.1 Time delay system and its description. The time domain representation of systems with delay L in the input portion is

$$\dot{x}(t) = Ax(t) + bu(t-L) \tag{10.1a}$$
$$y(t) = c^T x(t). \tag{10.1b}$$

On the other hand, if the delay is with output portion, it becomes

$$\dot{x}(t) = Ax(t) + bu(t) \tag{10.2a}$$
$$y(t) = c^T x(t-L). \tag{10.2b}$$

In either case, the transfer function is

$$t(s) \stackrel{d}{=} \frac{y(s)}{u(s)} = \frac{qr(s)}{p(s)} e^{-Ls}. \tag{10.3}$$

It is to be noticed that x(t) is only the state of lumped portion of the delay system. The whole state of the delay system (10.1) is $\{\ x(t);\ u(\tau),\ t-L\leq\tau\leq t\ \}$; that is, the initial conditions which uniquely determine the future free motion of the system.

10.2 Stabilizing control of delay system in the time domain. Before entering frequency domain approach, let us consider the pole assignment in the time domain. Especially, we consider the case when the plant with delay is unstable. It is well known that if the plant is controllable, arbitrary pole assignment is always possible by state variable feedback and therefore stabilizing control of unstable plant is possible. Since the state of the plant is $\{x(t);u(\tau)\}$, the state variable feedback assumes naturally the following form:

$$u(t) = f^T x(t) + \int_{-L}^{0} \lambda(\sigma) u(t+\sigma)\, d\sigma + v(t). \qquad (10.4)$$

The appropriate feedback f and $\lambda(\sigma)$ can be determined from eq.(10.1).

Suppose for the moment that u(t-L) could be determined by the equation

$$u(t-L) = f^T x(t) + v(t-L). \qquad (10.5)$$

Then arbitrary pole assignment will be achieved from the lumped parameter system theory. However, t-L is an instant in the past, and the value of u at that instant was already employed as the input. Therefore, eq.(10.5) cannot be used as a control law. If so, we shall rewrite eq.(10.5) as

$$u(t) = f^T x(t+L) + v(t). \qquad (10.6)$$

The signal x(t+L) can be calculated as

$$x(t+L) = e^{AL} x(t) + \int_{t}^{t+L} e^{A(t+L-\tau)} b u(\tau-L)\, d\tau. \qquad (10.7)$$

Substituting eq.(10.7) into eq.(10.6), we obtain

$$u(t) = f^T e^{AL} x(t) + \int_{t}^{t+L} f^T e^{A(t+L-\tau)} b u(\tau-L)\, d\tau + v(t)$$

$$= f^t e^{AL} x(t) + \int_{-L}^{0} f^T e^{-A\sigma} b u(t+\sigma)\, d\sigma + v(t), \qquad (10.8)$$

which assumes just the form of eq.(10.4), and hence is considered to provide the true state feedback. In practice, the inaccessible x(t) should be estimated. If $\{c^T, A\}$ is observable, x(t) can be estimated by Luenberger observer with u(t-L) and y(t) being input and output respectively. The finite poleassignment, the meaning of which is ex-plained in the next section, is achieved by the control law (10.8) [1],[2]

10.3 Poles of delay systems. Clearly, the zeros of p(s) are the poles of the delay system (10.3). Also, the delay system has infinite number of poles arising from e^{-Ls}. The former poles are called finite poles, while the latter poles are called infinite poles. The infinite poles lei on the straight line parallel to and infinitely leftward far from the imaginary axis. The imaginary parts of infinite poles assume the value $(2k+1)\pi/L$, $k=0, \pm 1, \pm 2, \cdots$.

Conventional feedback to control the plant moves all finite as well as infinite poles. Especially, feedback which moves unstable finite poles into LHP has a tendency to move some of infinite poles into RHP, resulting an unstable control system. The control law specified by eq.(10.8) provides arbitrary pole assignment for all finite poles while avoiding any movement of infinite poles which is considered as an ideal method of stabilizing control of delay system and is called finite pole assignment. The frequency domain approach quite equivalent to the above is presented in Section 10.4, which is further extended to the frequency domain exact model matching in Section 10.5.

10.4 Frequency domain finite pole assignment.[3] The plant transfer function is given by eq.(10.3), where r(s) and p(s) are m and n degree monic polynomials respectively with $0 \leq m \leq n-1$. It is assumed that r(s) and p(s) are relatively prime. For the mere simplicity of explanation $p(s)$ is assumed to have no multiple zeros.

Let the desired closed loop characteristic polynomial be $p_f(s)$, n degree monic polynomial, and put $f(s)=p(s)-p_f(s)$. Let the partial expansion of $f(s)/p(s)$ be

$$\frac{f(s)}{p(s)} = \sum_{i=1}^{n} \frac{\alpha_i}{s-\lambda_i} . \tag{10.9}$$

Then we define an n-1 degree polynomial $f_L(s)$ by

$$\frac{f_L(s)}{p(s)} = \sum_{i=1}^{n} \frac{\alpha_i e^{\lambda_i L}}{s-\lambda_i} . \tag{10.10}$$

Consider then the polynomial equation

$$k(s)p(s) + h(s)gr(s) = q(s)f_L(s), \tag{10.11}$$

where $q(s)$ is any n-1 degree monic stable polynomial, which yield unique solution for $k(s)$ amd $h(s)$ with of degree n-2 and n-1 respectively. Using the solutions $k(s)$ and $h(s)$, we consider a control law

$$u(t) = \frac{k(p)}{q(p)}u(t-L) + \frac{h(p)}{q(p)}y(t) + \int_{-L}^{0} \sum_{i=1}^{n} \alpha_i e^{-\lambda_i \sigma} u(t+\sigma)\, d\sigma + v(t). \tag{10.12}$$

Theorem 10.1 The control law (10.12) is realizable and achieves the desired pole assignment.

(proof) Both $k(p)q^{-1}(p)$ and $h(p)q^{-1}(p)$ are proper, and the signal $u(t+\sigma)$, $-L\leq\sigma\leq 0$, is definite. Thus the control law is realizable. The Laplace form of eq.(10.12) is

$$u(s) = \frac{k(s)}{q(s)}u(s)e^{-Ls} + \frac{h(s)}{q(s)}y(s)$$

$$+ \int_{-L}^{0} \sum_{i=1}^{n} \alpha_i e^{-\lambda_i \sigma} u(s) e^{\sigma s} \, d\sigma + v(s). \qquad (10.13)$$

However,

$$\int_{-L}^{0} \sum_{i=1}^{n} \alpha_i e^{-\lambda_i \sigma} u(s) e^{\sigma s} \, d\sigma$$

$$= \sum_{i=1}^{n} \alpha_i \int_{-L}^{0} e^{(s-\lambda_i)\sigma} \, d\sigma \cdot u(s)$$

$$= \sum_{i=1}^{n} \frac{\alpha_i}{s-\lambda_i} u(s) - \sum_{i=1}^{n} \frac{\alpha_i e^{\lambda_i L} e^{-Ls}}{s-\lambda_i} u(s)$$

$$= \frac{f(s)}{q(s)} u(s) - \frac{f_L(s)}{q(s)} u(s) e^{-Ls}.$$

Then,

$$u(s) = \frac{k(s)}{q(s)} u(s) e^{-Ls} + \frac{h(s)}{q(s)} y(s) + \frac{f(s)}{p(s)} u(s)$$

$$- \frac{f_L(s)}{p(s)} u(s) e^{-Ls} + v(s). \qquad (10.14)$$

On the other hand, from eq.(10.11), we obtain

$$\frac{k(s)}{q(s)} u(s) e^{-Ls} + \frac{h(s)}{q(s)} y(s) = \frac{f_L(s)}{p(s)} u(s) e^{-Ls}. \qquad (10.15)$$

From eqs.((10.4) and (10.15),

$$u(s) = \frac{f(s)}{p(s)} u(s) + v(s),$$

or

$$\frac{1}{p(s)} u(s) = \frac{1}{p_f(s)} v(s).$$

Therefore,

$$y(s) = \frac{gr(s)}{p(s)} u(s) e^{-Ls} = \frac{gr(s)}{p_f(s)} e^{-Ls} v(s). \qquad (10.16)$$

Q.E.D.

10.5 Frequency domain exact model matching.[3] A stable reference model transfer function is given by

$$t_d(s) \triangleq \frac{y_m(s)}{v(s)} = \frac{g_d r_d(s)}{p_d(s)} e^{-Ls}. \tag{10.17}$$

where $r_d(s)$ and $p_d(s)$ are m_d and n_d degree monic polynomials respectively. Two assumptions are made as follows.

A.1 $r(s)$ is stable.

A.2 $n_d - m_d \geq n - m$

We introduce any m and n degree monic stable polynomials $r^*(s)$ and $p^*(s)$ respectively to constitute a system

$$t_{IN} = \frac{r_d(s)p^*(s)}{p^*(s)r^*(s)}. \tag{10.18}$$

Clearly $t_{IN}(s)$ is proper and stable, we employ $t_{IN}(s)$ as an input dynamics. Denote $t_{IN}(s)v(s)$ by $\bar{v}(s)$.

We insert a static element $1/g$ before the plant, and regard the combined system {1/g + plant }, i.e. $r(s)p^{-1}(s)$ as a new plant. The input variable to the new plant is denoted by $u_b(t)$, which is a input signal to be synthesized.

Both the polynomials $r^*(s)p(s)$ and $r(s)p^*(s)$ are n+m degree monic polynomials, and hence $r^*(s)p(s)-r(s)p^*(s)$ is an n+m-1 degree polynomial. Then the partial fraction of $\{r^*(s)p(s)-r(s)p^*(s)\}/r^*(s)p(s)$ can be represented as

$$\frac{r^*(s)p(s)-r(s)p^*(s)}{r^*(s)p(s)} = \sum_{i=1}^{n+m} \frac{\beta_i}{s-z_i} . \tag{10.19}$$

We can then define an n+m degree monic polynomial $\phi(s)$ by

$$\frac{r^*(s)p(s)-\phi(s)}{r^*(s)p(s)} = \sum_{i=1}^{n+m} \frac{\beta_i e^{z_i L}}{s-z_i} . \qquad (10.20)$$

Now we set a polynomial equation for $k_b(s)$ and $h_b(s)$:

$$k_b(s)p(s)+h_b(s)r(s) = \gamma(s)r^*(s)p(s)-\gamma(s)\phi(s), \qquad (10.21)$$

where $\gamma(s)$ is any n-m-1 degree monic stable polynomial. It is known that eq.(10.20) yields unique solution $k_b(s)$ and $h_b(s)$ of degree n-2 and n-1 respectively. Using $k_b(s)$ and $h_b(s)$ thus determined, we consider a control law:

$$u_b(t) = \frac{k_b(p)}{\gamma(p)r^*(p)}u_b(t-L) + \frac{h_b(p)}{\gamma(p)r^*(p)}y(t)$$
$$+ \int_{-L}^{0} \sum_{i=1}^{n+m} \beta_i e^{-z_i\sigma} u_b(t+\sigma)\, d\sigma + g_d\bar{v}(t). \quad (10.22)$$

Theorem 10.2 The control law (10.22) is realizable and achieves desired exact model matching.

(proof) Realizability is evident. The Laplace transform of eq.(10.21) is

$$u_b(s) = \frac{k_b(s)}{\gamma(s)r^*(s)}u_b(s)e^{-Ls} + \frac{h_b(s)}{\gamma(s)r^*(s)}y(s)$$
$$+ \int_{-L}^{0} \sum_{i=1}^{n+m} \beta_i e^{-z_i\sigma} u_b(s)e^{\sigma s}\, d\sigma + g_d\bar{v}(s) \qquad (10.23)$$

However,

$$\int_{-L}^{0} \sum_{i=1}^{n+m} \beta_i e^{-z_i\sigma} u_b(s)e^{\sigma s}\, d\sigma$$
$$= \frac{r^*(s)p(s)-r(s)p^*(s)}{r^*(s)p(s)}u_b(s) - \frac{r^*(s)p(s)-\phi(s)}{r^*(s)p(s)} u_b(s)e^{-Ls}.$$

Then,

$$u_b(s) = \frac{k_b(s)}{\gamma(s)r^*(s)}u_b(s)e^{-Ls} + \frac{h_b(s)}{\gamma(s)r^*(s)}y(s)$$

$$+ \frac{r^*(s)p(s)-r(s)p^*(s)}{r^*(s)p(s)}u_b(s)$$

$$- \frac{r^*(s)p(s)-\phi(s)}{r^*(s)p(s)}u_b(s)e^{-Ls} + g_d\bar{\upsilon}(s). \qquad (10.24)$$

On the other hand, from eq.(10.20), we obtain

$$\frac{k_b(s)}{\gamma(s)r^*(s)}u_b(s)e^{-Ls} + \frac{h_b(s)}{\gamma(s)r^*(s)} = u_b(s)e^{-Ls} - \frac{\phi(s)}{r^*(s)p(s)}u_b(s)e^{-Ls}. \qquad (10.25)$$

From eqs.(10.24) and (10.25),

$$u_b(s) = u_b(s)e^{-Ls} + u_b(s) - \frac{r(s)p^*(s)}{r^*(s)p(s)}u_b(s) - u_b(s)e^{-Ls} + g_d\bar{\upsilon}(s),$$

or

$$\frac{r(s)}{p(s)}u_b(s) = \frac{g_d r^*(s)}{p^*(s)}\bar{\upsilon}(s).$$

Therefore,

$$y(s) = \frac{r(s)}{p(s)}u_b(s)e^{-Ls} = \frac{g_d r^*(s)}{p^*(s)}e^{-Ls}\,\bar{\upsilon}(s). \qquad (10.26)$$

Furthermore, we obtain

$$u_b(s) = \frac{g_d p(s)r^*(s)}{r(s)p^*(s)}\,\bar{\upsilon}(s). \qquad (10.27)$$

Since $r(s)$ is stable, $u_b(t)$ is bounded for any bounded $\bar{\upsilon}(t)$, and hence all signals within the control system are bounded. Q.E.D.

The control law (10.22) can be written in terms of $u(t)$:

$$u(t) = \frac{1}{g}\{\frac{k(p)}{\gamma(p)r^*(p)}u(t-L) + \frac{h(p)}{\gamma(p)r^*(p)}y(t)$$

$$+ \int_{-L}^{0} \sum_{i=1}^{n+m} g\beta_i e^{-z_i\sigma} u(t+\sigma)\, d\sigma \} + \frac{g_d}{g} \bar{v}(t), \quad (10.28)$$

where $k(p)=gk_b(p)$ and $h(p)=h_b(p)$.

10.6 Adaptive control. Exact model matching technique can easily be extended to adaptive control. The control law when the plant is known is given by eq.(10.28). Let us define 2n-1 dimensional vectors θ and $\omega(t)$ by

$$\theta = [-k_{n-2}, \cdots, -k_0, -h_{n-1}, \cdots, -h_0]^T, \quad (10.29)$$

and

$$\omega(t) = [\frac{p^{n-2}}{T(p)r^*(p)}u(t-L), \cdots, \frac{1}{T(p)r^*(p)}u(t-L), \frac{p^{n-1}}{T(p)r^*(p)}y(t), \cdots, \frac{1}{T(p)r^*(p)}y(t)]^T, \quad (10.30)$$

where k_i and h_i are the coefficients of polynomials $k(s)$ and $h(s)$ respectively; i.e.,

$$k(s) = k_{n-2}s^{n-2} + \cdots + k_0, \quad (10.31)$$

and

$$h(s) = h_{n-1}s^{n-1} + \cdots + h_0. \quad (10.32)$$

Define further a time function $\lambda(\sigma)$ by

$$\lambda(\sigma) = - \sum_{i=1}^{n+m} g\, \beta_i\, e^{-z_i\sigma}. \quad (10.33)$$

Then, the control law (10.28) can be written as

$$u(t) = \frac{1}{g}\{-\theta^T\omega(t) - \int_{-L}^{0} \lambda(\sigma)u(t+\sigma)\, d\sigma\} + \frac{g_d}{g} \bar{v}(t). \quad (10.34)$$

It is to be noticed that $\int_{-L}^{0} \lambda(\sigma)u(t+\sigma)\, d\sigma$ is an inner product of infinite dimensional parameter $\lambda(\sigma)$ and infinite dimensional signal $u(t+\sigma)$. Also, notice that when L is zero, eq.(10.34) reduces to the finite dimensional control law (4.35).

Now, g, θ, and $\lambda(\sigma)$ are unknown in the adaptive case. Then, the control law is modified to the following, by replacing g, θ, and $\lambda(\sigma)$ by their respective estimates $\tilde{g}(t)$, $\tilde{\theta}(t)$ and $\tilde{\lambda}(t,\sigma)$ which will be defined formerly later. That is,

$$u(t) = \frac{1}{\tilde{g}(t)}\{-\tilde{\theta}^T(t)\omega(t) - \int_{-L}^{0} \tilde{\lambda}(t,\sigma)u(t+\sigma)\,d\sigma\} + \frac{g_d}{\tilde{g}(t)}\,\bar{v}(t). \qquad (10.35)$$

Like the finite dimensional adaptive case, we need an alternative representation for the plant dynamics. From eq.(10.21) we obtain

$$\frac{k_b(s)}{\tau(s)p^*(s)}u_b(s)e^{-Ls} + \frac{h_b(s)}{\tau(s)p^*(s)}y(s) = \frac{r^*(s)}{p^*(s)}u_b(s)e^{-Ls} - \frac{\phi(s)}{p(s)p^*(s)}u_b(s)e^{-Ls}. \qquad (10.36)$$

However,

$$\frac{\phi(s)}{p(s)p^*(s)}u_b(s)e^{-Ls} = \frac{\phi(s)}{r^*(s)p(s)}\cdot\frac{r^*(s)}{p^*(s)}u_b(s)e^{-Ls}$$

$$= \{1 - \sum_{i=1}^{n+m} \frac{\beta_i e^{z_i L}}{s-z_i}\} \frac{r^*(s)}{p^*(s)}u_b(s)e^{-Ls}.$$

Therefore,

R.H.S. of eq.(10.36)

$$= \sum_{i=1}^{n+m} \frac{\beta_i e^{z_i L}}{s-z_i} \frac{r^*(s)}{p^*(s)}u_b(s)e^{-Ls}$$

$$= -\{\sum_{i=1}^{n+m} \frac{\beta_i}{s-z_i} \frac{r^*(s)}{p^*(s)}u_b(s) - \sum_{i=1}^{n+m} \frac{\beta_i e^{z_i L}}{s-z_i} \frac{r^*(s)}{p^*(s)}u_b(s)e^{-Ls}\}$$

$$+\sum_{i=1}^{n+m} \frac{\beta_i}{s-z_i} \frac{r^*(s)}{p^*(s)} u_b(s)$$

$$= -\left\{\sum_{i=1}^{n+m} \frac{\beta_i}{s-z_i} \frac{r^*(s)}{p^*(s)} u_b(s) - \sum_{i=1}^{n+m} \frac{\beta_i e^{z_i L}}{s-z_i} \frac{r^*(s)}{p^*(s)} u_b(s) e^{-Ls}\right\}$$

$$+\frac{r^*(s)}{p^*(s)} u_b(s) - \frac{r(s)}{p(s)} u_b(s).$$

Thus, eq.(10.36) can be written as

$$\frac{k_b(s)}{\gamma(s)r^*(s)} \frac{r^*(s)}{p^*(s)} u_b(s) e^{-Ls} + \frac{h_b(s)}{\gamma(s)r^*(s)} \frac{r^*(s)}{p^*(s)} y(s)$$

$$= -\left\{\sum_{i=1}^{n+m} \frac{\beta_i}{s-z_i} \frac{r^*)s)}{p^*(s)} u_b(s) - \sum_{i=1}^{n+m} \frac{\beta_i e^{z_i L}}{s-z_i} \frac{r^*(s)}{p^*(s)} u_b(s) e^{-Ls}\right\}$$

$$+ \frac{r^*(s)}{p^*(s)} u_b(s) - \frac{r(s)}{p(s)} u_b(s). \qquad (10.37)$$

The time domain representation of eq.(10.37) is

$$\frac{k_b(p)}{\gamma(p)r^*(p)} \frac{r^*(p)}{p^*(p)} u_b(t-L) + \frac{h_b(p)}{\gamma(p)r^*(p)} \frac{r^*(p)}{p^*(p)} y(t)$$

$$= -\int_{-L}^{0} \sum_{i=1}^{n+m} \beta_i \, e^{-z_i\sigma} \frac{r^*(p)}{p^*(p)} u_b(t+\sigma)\, d\sigma$$

$$+ \frac{r^*(p)}{p^*(p)} u_b(t) - y(t+L),$$

which will further be rewritten into

$$\frac{k(p)}{\gamma(p)r^*(p)} \frac{r^*(p)}{p^*(p)} u(t-L) + \frac{h(p)}{\gamma(p)r^*(p)} \frac{r^*(p)}{p^*(p)} y(t)$$

$$= -\int_{-L}^{0} \sum_{i=1}^{n+m} g\beta_i e^{-z_i\sigma} \frac{r^*(p)}{p^*(p)} u(t+\sigma)\, d\sigma + g\frac{r^*(p)}{p^*(p)} u(t)$$

$$- y(t+L), \qquad (10.38)$$

by using $u(t)$, $k(t)$, and $h(t)$ instead of $u_b(t)$, $k_b(t)$, and $h_b(t)$ respectively. By shifting the time origin by $-L$, we obtain

$$y(t) = g\frac{r^*(p)}{p^*(p)}u(t-L) + \frac{k(p)}{\gamma(p)r^*(p)}\,\frac{r^*(p)}{p^*(p)}u(t-2L)$$
$$+ \frac{h(p)}{\gamma(p)r^*(p)}\,\frac{r^*(p)}{p^*(p)}y(t-L)$$
$$- \int_{-L}^{0} \sum_{i=1}^{n+m} g\beta_i e^{-z_i L}\,\frac{r^*(p)}{p^*(p)}u(t+\sigma-L)\,d\sigma. \quad (10.39)$$

Finally, using notations (10.29), (10,30), and (10.33), we obtain

$$y(t)=g\frac{r^*(p)}{p^*(p)}u(t-L) + \theta^T\frac{r^*(p)}{p^*(p)}\omega(t-L) + \int_{-L}^{0} \lambda(\sigma)\frac{r^*(p)}{p^*(p)}$$
$$\times u(t+\sigma-L)\,d\sigma, \quad (10.40)$$

as the representation of the plant dynamics in terms of g, θ, and $\lambda(\sigma)$. Notice that when L is equal to zero, eq.(10.40) reduces to eq.(4.37), the finite dimensional plant dynamics in terms of g and θ.

The adaptive control error e(t) is defined by $e(t) = y(t) - y_m(t)$; i.e.,

$$e(t) = g\frac{r^*(p)}{p^*(p)}u(t-L) + \theta^T\frac{r^*(p)}{p^*(p)}\omega(t-L)$$
$$+ \int_{-L}^{0} \lambda(\sigma)\frac{r^*(p)}{p^*(p)}u(t+\sigma-L)\,d\sigma - y_m(t), \quad (10.41)$$

which represents a dynamics of the error system. The adaptive law is derived by identifying the error system. The identifier is defined by using estimates $\tilde{g}(t)$, $\tilde{\theta}(t)$, and $\tilde{\lambda}(t,\sigma)$; i.e.,

$$\tilde{e}(t) = \tilde{g}(t)\frac{r^*(p)}{p^*(p)}u(t-L) + \tilde{\theta}^T(t)\frac{r^*(p)}{p^*(p)}\omega(t-L)$$
$$+ \int_{-L}^{0} \tilde{\lambda}(t,\sigma)\frac{r^*(p)}{p^*(p)}u(t+\sigma-L)\,d\sigma - y_m(t). \quad (10.42)$$

The identification error $\varepsilon(t)$ is defined by $\varepsilon(t)=\tilde{e}(t) - e(t)$. That is,

$$\varepsilon(t) = [\tilde{g}(t)-g]\,\frac{r^*(p)}{p^*(p)}u(t-L)$$
$$+ [\tilde{\theta}(t)-\theta]^T\,\frac{r^*(p)}{}\omega(t-L)$$

$$+\int_{-L}^{0} [\tilde{\lambda}(t,\sigma)-\lambda]\, \frac{r^*(p)}{p^*(p)}u(t+\sigma-L)\, d\sigma. \tag{10.43}$$

Notice that eq.(10.43) is a natural extension of eq. (4.31), the identification error for finite dimensional adaptive control.

Let us define $\Omega(t)$ by

$$\Omega(t) = [\frac{r^*(p)}{p^*(p)}u(t-L) \quad \frac{r^*(p)}{p^*(p)}\omega^T(t-L) \quad \frac{r^*(p)}{p^*(p)}u(t+\sigma-L)]^T, \tag{10.44}$$

and $||\Omega(t)||^2$ by

$$||\Omega(t)||^2 = [\frac{r^*(p)}{p^*(p)}u(t-L)]^2 + [\frac{r^*(p)}{p^*(p)}\omega^T(t-L)] \times [\frac{r^*(p)}{p^*(p)}\omega(t-L)] + \int_{-L}^{0} [\frac{r^*(p)}{p^*(p)}u(t+\sigma-L)]^2\, d\sigma. \tag{10.45}$$

Then, the adaptive law can be given by

$$\left.\begin{aligned}
\dot{\tilde{g}}(t) &= -\tau_g\, \frac{r^*(p)}{p^*(p)}u(t-L)/\{c+||\Omega(t)||^2\},\quad \tau_g>0 \\
\dot{\tilde{\theta}}(t) &= -\Gamma_\theta\, \frac{r^*(p)}{p^*(p)}\omega(t-L)/\{c+||\Omega(t)||^2\},\quad \Gamma_\theta=\Gamma_\theta^T>0 \\
\dot{\tilde{\lambda}}(t,\sigma) &= -\tau_\lambda(\sigma)\frac{r^*(p)}{p^*(p)}u(t+\sigma-L)/\{c+||\Omega(t)||^2,\quad \tau_\lambda(\sigma)>0 \\
& \qquad c>0
\end{aligned}\right\}. \tag{10.46}$$

The stability analysis is quite similar to that for finite dimensional adaptive control, and is omitted here.

REFERENCES

[1] A.Z.Manitus and A.W.Olbrot; Finite spectrum assignment problem for system with delay, IEEE Vol.AC-24,541/553, 1979.

[2] T.Furukaw and E.Shimemura; Predictive control for systems with time delay, Int.J.Control,37,399/412,1983.

[3] K.Ichikawa; Frequency domain pole assignment and exact model matching for delay systems, to appear in Int. J. Control.

Lecture Notes in Control and Information Sciences

Edited by M. Thoma

Vol. 43: Stochastic Differential Systems
Proceedings of the 2nd Bad Honnef Conference of the SFB 72 of the DFG at the University of Bonn June 28 – July 2, 1982
Edited by M. Kohlmann and N. Christopeit
XII, 377 pages. 1982.

Vol. 44: Analysis and Optimization of Systems
Proceedings of the Fifth International Conference on Analysis and Optimization of Systems Versailles. December 14–17, 1982
Edited by A. Bensoussan and J. L. Lions
XV, 987 pages, 1982

Vol. 45: M. Arató
Linear Stochastic Systems with Constant Coefficients
A Statistical Approach
IX, 309 pages. 1982

Vol. 46: Time-Scale Modeling of Dynamic Networks with Applications to Power Systems
Edited by J. H. Chow
X, 218 pages. 1982

Vol. 47: P. A. Ioannou, P. V. Kokotovic
Adaptive Systems with Reduced Models
V, 162 pages. 1983

Vol. 48: Yaakov Yavin
Feedback Strategies for Partially Observable Stochastic Systems
VI, 233 pages, 1983

Vol. 49: Theory and Application of Random Fields
Proceedings of the IFIP-WG 7/1 Working Conference held under the joint auspices of the Indian Statistical Institute
Bangalore, India, January 1982
Edited by G. Kallianpur
VI. 290 pages. 1983

Vol. 50: M. Papageorgiou
Applications of Automatic Control Concepts to Traffic Flow Modeling and Control
IX, 186 pages. 1983

Vol. 51: Z. Nahorski, H. F. Ravn, R. V. V. Vidal
Optimization of Discrete Time Systems
The Upper Boundary Approach
V, 137 pages 1983

Vol. 52: A. L. Dontchev
Perturbations, Approximations and Sensitivity Analysis of Optimal Control Systems
IV, 158 pages. 1983

Vol. 53: Liu Chen Hui
General Decoupling Theory of Multivariable Process Control Systems
XI, 474 pages. 1983

Vol. 54: Control Theory for Distributed Parameter Systems and Applications
Edited by F. Kappel, K. Kunisch, W. Schappacher
VII, 245 pages. 1983.

Vol. 55: Ganti Prasada Rao
Piecewise Constant Orthogonal Functions and Their Application to Systems and Control
VII, 254 pages. 1983.

Vol. 56: Dines Chandra Saha, Ganti Prasada Rao
Identification of Continuous Dynamical Systems
The Poisson Moment Functional (PMF) Approach
IX, 158 pages. 1983.

Vol. 57: T. Söderström, P. G. Stoica
Instrumental Variable Methods for System Identification
VII, 243 pages. 1983.

Vol. 58: Mathematical Theory of Networks and Systems
Proceedings of the MTNS-83 International Symposium
Beer Sheva, Israel, June 20–24, 1983
Edited by P. A. Fuhrmann
X, 906 pages. 1984

Vol. 59: System Modelling and Optimization
Proceedings of the 11th IFIP Conference
Copenhagen, Denmark, July 25-29, 1983
Edited by P. Thoft-Christensen
IX, 892 pages. 1984

Vol. 60: Modelling and Performance Evaluation Methodology
Proceedings of the International Seminar
Paris, France, January 24–26, 1983
Edited by F. Bacelli and G. Fayolle
VII, 655 pages. 1984

Vol. 61: Filtering and Control of Random Processes
Proceedings of the E.N.S.T.-C.N.E.T. Colloquium
Paris, France, February 23–24, 1983
Edited by H. Korezlioglu, G. Mazziotto, and J. Szpirglas
V, 325 pages. 1984

Lecture Notes in Control and Information Sciences

Edited by M. Thoma

Vol. 62: Analysis and Optimization of Systems
Proceedings of the Sixth International Conference on Analysis and Optimization of Systems
Nice, June 19–22, 1984
Edited by A. Bensoussan, J. L. Lions
XIX, 591 pages. 1984.

Vol. 63: Analysis and Optimization of Systems
Proceedings of the Sixth International Conference on Analysis and Optimization of Systems
Nice, June 19–22, 1984
Edited by A. Bensoussan, J. L. Lions
XIX, 700 pages. 1984.

Vol. 64: Arunabha Bagchi
Stackelberg Differential Games in Economic Models
VIII, 203 pages, 1984

Vol. 65: Yaakov Yavin
Numerical Studies in Nonlinear Filtering
VIII, 273 pages, 1985.

Vol. 66: Systems and Optimization
Proceedings of the Twente Workshop
Enschede, The Netherlands, April 16–18, 1984
Edited by A. Bagchi, H. Th. Jongen
X, 206 pages, 1985.

Vol. 67: Real Time Control of Large Scale Systems
Proceedings of the First European Workshop
University of Patras, Greece, Juli 9–12, 1984
Edited by G. Schmidt, M. Singh, A. Titli, S. Tzafestas
XI, 650 pages, 1985.

Vol. 68: T. Kaczorek
Two-Dimensional Linear Systems
IX, 397 pages, 1985.

Vol. 69: Stochastic Differential Systems – Filtering and Control
Proceedings of the IFIP-WG 7/1 Working Conference
Marseille-Luminy, France, March 12-17, 1984
Edited by M. Metivier, E. Pardoux
X, 310 pages, 1985.

Vol. 70: Uncertainty and Control
Proceedings of a DFVLR International Colloquium
Bonn, Germany, March, 1985
Edited by J. Ackermann
IV, 236 pages, 1985.

Vol. 71: N. Baba
New Topics in Learning Automata Theory and Applications
VII, 231 pages, 1985.

Vol. 72: A. Isidori
Nonlinear Control Systems: An Introduction
VI, 297 pages, 1985.

Vol. 73: J. Zarzycki
Nonlinear Prediction Ladder-Filters for Higher-Order Stochastic Sequences
V, 132 pages, 1985.

Vol. 74: K. Ichikawa
Control System Design based on Exact Model Matching Techniques
VII, 129 pages, 1985.

GPSR Compliance
The European Union's (EU) General Product Safety Regulation (GPSR) is a set of rules that requires consumer products to be safe and our obligations to ensure this.

If you have any concerns about our products, you can contact us on

ProductSafety@springernature.com

In case Publisher is established outside the EU, the EU authorized representative is:

Springer Nature Customer Service Center GmbH
Europaplatz 3
69115 Heidelberg, Germany

www.ingramcontent.com/pod-product-compliance
Ingram Content Group UK Ltd.
Pitfield, Milton Keynes, MK11 3LW, UK
UKHW052004200726
13877UKWH00031BA/691
9783662166406